FOUNDATIONS FOR NANOSCIENCE AND NANOTECHNOLOGY

FOUNDATIONS FOR NANOSCIENCE AND NANOTECHNOLOGY

Nils O. Petersen

CRC Press
Taylor & Francis Group
Boca Raton London New York

CRC Press is an imprint of the
Taylor & Francis Group, an **informa** business

CRC Press
Taylor & Francis Group
6000 Broken Sound Parkway NW, Suite 300
Boca Raton, FL 33487-2742

© 2017 by Taylor & Francis Group, LLC
CRC Press is an imprint of Taylor & Francis Group, an Informa business

No claim to original U.S. Government works

Printed on acid-free paper
Version Date: 20170303

International Standard Book Number-13: 978-1-4822-5907-0 (Hardback)

Library of Congress Cataloging-in-Publication Data

Names: Petersen, Nils O., author.
Title: Foundations for nanoscience and nanotechnology / Nils O. Petersen.
Description: Boca Raton, FL : CRC Press, Taylor & Francis Group, [2017] |
Includes bibliographical references and index.
Identifiers: LCCN 2016049574| ISBN 9781482259070 (pbk. ; alk. paper) | ISBN
1482259079 (pbk. ; alk. paper) | ISBN 9781482259100 (e-book) | ISBN
1482259109 (e-book) | ISBN 9781315321370 (e-book) | ISBN 1315321378
(e-book) | ISBN 9781482259087 (e-book) | ISBN 1482259087 (e-book)
Subjects: LCSH: Nanoscience. | Nanotechnology.
Classification: LCC QC176.8.N35 P47 2017 | DDC 500--dc23
LC record available at https://lccn.loc.gov/2016049574

Visit the Taylor & Francis Web site at
http://www.taylorandfrancis.com

and the CRC Press Web site at
http://www.crcpress.com

Printed and bound in the United States of America by
Edwards Brothers Malloy on sustainably sourced paper

This book is dedicated to all those brave souls who seek to understand the fascinating world of the really small.

Contents

PART IV Understanding fluctuations in small systems

Preface

This books title emphasizes **Foundations for** since it is not intended to be a book **about** nanoscience or nanotechnology and the numerous applications arising from the rapid research and development in these very exciting fields. Rather, it is a book aimed at providing some of the principles of physics and chemistry that are needed to understand why the science of small systems are different than those of bulk systems.

This book is written primarily as a textbook for senior undergraduate students and beginning graduate students interested in understanding how the unique properties of nanoscale materials and systems arise. The book aims to have enough detail to be useful for courses that introduce students to the basics of physical chemistry but is not specifically targeted at students from chemistry or physics who already have advanced courses at the undergraduate or graduate levels in quantum mechanics, thermodynamics, statistical mechanics, or spectroscopy. These students may, however, find that parts of this book provide useful reference material for their work. **The primary target audience** includes those students who are introduced to or work in the fields of nanoscience and nanotechnology but come at it from diverse backgrounds such as material sciences, biology, or engineering.

The philosophy of the book is to provide a fairly rigorous treatment of the topics at the simplest level in order to provide the reader with the essential principles and a general understanding of the foundations for key observations, such as the effect of the dimension of the nanoscale system. The book then draws on that foundation to illustrate how these principles can be applied to more complex systems that may better approximate real systems without necessarily providing the detailed solutions. The book finally provides selected examples from the literature that illustrate how the foundational knowledge can be used to understand, at least in principle, experimental observations and that introduces the readers to some recent applications of nanoscience and nanotechnology. With this approach, some of the early parts of every section of each chapter may be too introductory for some readers, and they may be easily skipped or serve as a reminder.

While it is not the intent of this book to provide the detailed or rigorous approach to solving the physical chemical properties in complex systems, those students or researchers who need to go to that next level, should find this book a useful first step toward that objective.

The topics covered in this textbook have been selected to, on the one hand, cover the most important foundations for understanding nanoscience and nanotechnology, such as the effect of quantum confinement (Part II, Chapters 2-5) and the impact on surface properties (Part III, Chapters 6-9); and on the other hand, to introduce topics that are not usually covered in depth, such as the importance of fluctuations in small systems at the nanoscale (Part IV, Chapters 10-13) and the interaction of light with matter (Part V Chapter 14-17). On completion of this book, students should be able to read much of the modern literature in nanoscience and nanotechnology with a good understanding of the topics and their importance. The book is, I believe, unique by bringing together in one place a number of topics that frequently are covered separately.

At the end of the book, I have included a number of problems that hopefully will help students better understand the material. Some problems are simply aimed to assist in remembering key observations or conclusions, others are designed to expand on the subject, and yet others are created for the more advanced student who is interested in working through the details of some of the derivations of key equations.

It should be noted that the book is restricted to subjects with which I have some experience or some understanding. There are a number of topics which are deliberately not covered because I don't feel I have the expertise to cover them in sufficient detail to make them useful. These topics include magnetic and mechanical properties at the nanoscale as well as nanoelectronics beyond the very basics. These are clearly topics of great importance and if this text proves valuable, future versions may include such sections written by other contributing authors.

I

Introduction

CHAPTER 1

Why study the foundations for nanoscience and nanotechnology?

1.1 NANOSCIENCE VERSUS NANOTECHNOLOGY

So what are we talking about when we talk about nanoscience, nanotechnology, nanomaterials, nano-objects or anything else with the prefix nano (other than a unit)? It turns out to be hard to define in a way that captures all aspects of this world of the nanoscale.

At first glance, it seems simple: objects or materials that are on a scale measured by the unit of a nanometer (say 1-100 nm). And this is indeed one component of almost all definitions. However, that turns out not to be sufficiently precise to communicate either scientifically or in legal terms what exactly we are working with. In fact, since the emergence of the concepts of nanoscience and nanotechnology, numerous definitions have been developed in different jurisdictions across the world and as of this writing there is no universally accepted definition. Why is this important? The exact definition may affect how nanoscale materials or object are subjected to regulations of manufacture or use, whether they can be traded, or how they may be incorporated into products many things.

It may be that a unique definition is impossible, but it seems that a good definition needs to refer to the scale (1-100 nm); to the fact that it has unique properties as a consequence of the scale; and whether it is natural or man-made.

Some definitions of nanomaterials:

- A material with any external dimension in the nanoscale (the length ranging from approximately 1 nm to 100 nm) or having internal structure or surface structure in the nanoscale — *International Organization for Standarization (ISO)*

- A chemical that is either a nano-object (a material confined in one, two, or three dimensions at the nanoscale (the size typically range between 1 nm and 100 nm)) or is nano-structured (having an internal or surface structure at the nano scale) — *OECD*

- Nanomaterials can exhibit unique optical, mechanical, magnetic, conductive, and sorptive properties different than the same chemical substances in a large size — *US Environmental Protection Agency*

- Both materials that have at least one dimension in the size range of approximately 1 nanometers (nm) to 100 nm and certain materials that otherwise exhibit related dimension-dependent properties and phenomena — *US Food and Drug Administration*

- Materials at or within the nanoscale (meaning 1 to 100 nanometers, inclusive) in at least one dimension, or has internal or surface structures at the nanoscale; or it is smaller or larger than the nanoscale in all dimensions and exhibits one or more nanoscale properties/phenomena — *Health Canada*

- A natural, incidental, or manufactured material containing particles, in an unbound state or as an aggregate or as an agglomerate and where, for 50% or more of the particles in the number size distribution, one or more external dimensions is in the size range 1 nm - 100 nm — *European Union*

For the purpose of this text, I will argue that nanoscience refers to **exploring** the properties of materials at small scales while nanotechnology refers to **exploiting** our understanding of the properties of materials at the small scale. This to me is a useful distinction from the point of view of thinking about the evolution of research into innovation: the prerequisite for developing useful products or processes that are based on nanoscale phenomena is our understanding of the fundamental science underlying these phenomena.

The United States National Nanotechnology Initiative argues that the concept of nanotechnology encompasses nanoscale science, engineering, and technology where nanoscience is seeking to understand the fundamental properties and nanoengineering is seeking to effectively use these properties. In this context, they state that *"nanotechnology is the understanding and con-*

trol of matter at dimensions between approximately 1 and 100 nanometers where unique properties enable novel applications not feasible when working with bulk materials or even single atoms or molecules." Nanotechnology therefore includes visualizing, measuring, modeling, designing, and manipulating chemical matter at this small scale.

Whatever the exact use of the terminology, it is evident that for us to exploit nanoscale phenomena, we need to understand what is so different about matter at this scale.

One **foundation** for understanding the nanoscale work was created more than a century ago with the development of quantum mechanics. This is the theoretical framework that replaces classical Newtonian mechanics when the scale is small. Thus we have had a theoretical machinery to describe behaviors of atoms and electrons in molecules to explain concepts such as bonds to form molecules. This machinery is now able to explain many of the key properties of nanomaterials that arise from confining electrons and molecules to small spaces as we explore in Part II of this book.

A second **foundation** for understanding the properties of nanomaterials was established a few centuries ago when the field of thermodynamics emerged. While thermodynamics primarily provides the means of understanding the flow of energy through systems, it also provides insight into the energetics of surfaces, which become very important as the surface to volume ratio increases for small particles. The energetics of interactions between surfaces has also been explored for a long time in the field known as colloid science and this can be extended to the nanoscale as well. These well-known theoretical frameworks will be developed in Part III of this book.

A third **foundation** for understanding some peculiarities of small systems was also developed more than a century ago as Statistical Mechanics created a bridge between quantum mechanics and thermodynamics. The importance of understanding the statistical nature of properties in small systems is investigated in Part IV.

While not a foundation, in the same sense as those listed above, electromagnetic radiation (light) and its interaction with matter has proven a critical tool for studying nanoscale systems. Part V therefore explores various aspects of interaction of light with matter with particular emphasis on absorption and emission processes, which are easily observed even for very small systems or systems with only a few nanoscale materials.

So what is new about nanotechnology? It appears that we have known a lot about small systems for a long time and indeed we have. However, there are new developments that have emerged in two areas: visualizing and manipulating at a small scale. The ability to visualize matter at a small scale has steadily improved as tools such as x-ray diffraction and electron microscopy has developed to the point where we can determine structural properties of pieces of material as small as a few nanometers across. In some cases it is now even possible to study the structure and behavior of individual molecules. Most importantly, however, the last several decades has seen invention of new tools that in addition to visualizing individual atoms and molecules, have allowed us to manipulate these almost at will. This new element of control has fed the excitement around nanoscience and nanotechnology and laid the foundation for new design principles.

Furthermore, our understanding of biological processes has developed rapidly in this century and has revealed that much of life is based on systems that operate at the nanoscale, such as membranes, catalytic protein complexes, and synthetic and repair mechanisms. This new insight has provided inspiration for development of nanotechnologies that mimic biological processes in a number of ways.

Thus the emergence of nanotechnology as a topic of great interest arises from the convergence of long standing understanding of material properties, of new tools to visualize and manipulate individual atoms or molecules, and of new ways to design new structures or materials based on understanding of molecular biological systems. In addition, growth of the power to compute has enabled modeling and calculations of systems on larger and larger scales from atoms up to assemblies of molecules in nanoscale structures.

1.2   THE IMPORTANCE OF NANOSCALE PHENOMENA

This section will provide a brief overview of some of the phenomena that distinguish nanoscale properties from bulk properties and hence provides the motivation for the reader to want to understand the fundamental principles that form the foundation for nanoscience and nanotechnology

So why do nanoscale systems behave differently than bulk systems, on the one hand, and individual atoms or molecules, on the other hand? We know now that there are two fundamentally important phenomena that underlie these differences: The first is the effect of confining electrons to small dimensions, this is called **quantum confinement**. This leads to a host of changes in properties that depend on the size of the nano-object. The second is the effect of **increasing surface area** relative to the bulk meaning that the properties of atoms at the interface become more important than those in the interior of the material. In the extreme there is virtually no bulk material. Yet, for both effects, the assembly of atoms or molecules into small clusters or particles impart properties of the assembly that differ from the individual, constituent parts.

The consequence of both of these effects — quantum confinement and surface-to-volume ratio — is that **size matters**. And this understanding is key to the exploitation of the material properties of nanoscale objects. If we can control the size and shape of the objects, we can control the properties. We have the basis for design of objects with predictable, known properties, so we can create new materials to fulfill particular, desirable functions that solve specific problems of interest.

We can create nanoparticles with desirable catalytic properties for use in catalytic converters. We can create nanoporous materials that will exclusively bind specific molecules to, for example, capture noxious gases emitted at a power plant. We can create structured assemblies that facilitate the absorption of light and capture and transfer electrons into a circuit to enhance the efficiency of solar cells. The possibilities are only limited by our imagination, and that is what makes this such an exciting field for science, engineering, and technology development.

The rapidly evolving understanding of biological systems has inspired the design principles of nanotechnology researchers and the intersection of molecular biology and nanotechnology has been critical for the rapid development of the field. In addition, it turns out that the tools and techniques of nanotechnology has been instrumental in unraveling biological phenomena and the products of the nanotechnology revolution promise to have applications in the life sciences, particularly in the areas of diagnostics, therapeutics, and studies of biological structures, dynamics, and interactions. Some of these tools at the interface between biology and nanotechnology are explored in Part V of the book.

1.3   NANOSCIENCE AND NANOTECHNOLOGY FUTURE

In the initial phases of the discovery of the nanotechnology world, the excitement created great expectations that nanotechnology was the next phase of economic development, following the path of information technology and biotechnology. These expectations are slowly being met, but perhaps not at the rate hoped for in the early days. Still, there is strong evidence that the impact and applications of nanotechnology will be **pervasive, persistent,** and **powerful**.

Starting in 2005, the Woodrow Wilson International Center for Scholars and the Pew Charitable Trust started the project on Emerging Nanotechnologies to keep track of the development of nanotechnologies and to help ensure that the potential benefits of these technologies can be developed with an eye to continual public engagement to minimize potential risks of their introduction. Their web-site (http://www.nanotechproject.org/) has been and will remain a valuable resource for both the public and specialist.

One component of the project is creation of an ongoing inventory of consumer goods and other applications that use nanotechnology as a key component. During the first decade or so, this inventory has grown to nearly two thousand products across a large number of domains, supporting the notion of nanotechnology applications being pervasive, persistent, and powerful.

1.3.1   Pervasiveness

Because nanotechnology is first and foremost about creating and applying new materials with new properties from, for the most part, known chemical matter, it may be viewed as a special branch of material science. This means that we expect nanomaterials to become integral parts of products in just about every discipline or area that affects consumers and society, ranging from energy, transportation, manufacturing, and climate change to agriculture, health, communications, and food.

The Project of Emerging Nanotechnologies consumer products inventory lists the nanotechnology enabled products in a number of broad categories such as appliances, including those for kitchens and laundry; automotive; electronic and computers; food and beverage; health and fitness; and home and garden. In addition, they feature applications in agriculture, environmental remediation, and medicine. There are very few domains in which nanotechnology will not be or become an important component.

As an example, the areas of agriculture and food currently support research and development projects in areas as diverse as veterinary medicine, sustainable agriculture, bio-sensors, bio-processing for foods, pathogen detection, and environmental processing.

1.3.2   Persistence

The success of nanotechnology to date is based on the convergence of disciplines ranging from chemistry, biology, and physics to various branches of engineering, mathematics, computational sciences and business. As a consequence of the meeting of different minds, the approach to the nanosciences more often than not involve multidisciplinary teams that have changed how we think about conducting research in this domain. We now exploit our ability to **visualize** structure, function, and dynamics at the nanoscale with ever increasing precision; to integrate this information to **design** structures and materials with predictable and desirable properties; and to **control** the assembly of the parts or the machining of components at the molecular level. This holistic approach is broadening the scope of possible areas of application of nanoscience and means that the applications of nanotechnology will persist and evolve just as applications of electronics and computational paradigms persist and evolve. Once unearthed, nanoscience and nanotechnology will be with us forever as platforms for research and technology development.

That said, it is important to recognize that nanoscience and nanotechnology provide us with enabling technologies that will most likely be critical or integral components of other technologies, that are improved or are made to exist as a result of their presence. In some cases, the nanotechnologies will make products that are smaller or faster, in other cases that are easier to produce or cheaper, and in yet other cases that break new ground.

1.3.3   Power

The National Nanotechnology Initiative reported in 2010 on the first decade of evolution of nanotechnology with focus on the United States, but with global information included. This report, titled *Nanotechnology Research Directions for Societal Needs in 2020. Retrospective and Outlook* provided a thorough review of the progress in research as well as on the economic impact and provided projections and predictions about how the next decade would unfold.

There were a number of exciting findings which demonstrated how nanotechnology had the potential to become a powerful economic driver for the future. During nearly a decade, research investments, publications, patent applications, products market, and the workforce associated with nanotechnology grew by an astounding rate of about 25% per year and by 2009, the global value of nanotechnology enabled products had reached about US250B$. The report predicted that during the next decade, this growth would continue and reach a total value of US3T$.

Equally interesting was the analysis of how the nanotechnology products and processes were anticipated to evolve. Starting with applications of passive nanostructures, moving through active nanostructures to nanosystems to converging technologies, the field was projected to become more and more sophisticated in its application of the unique properties of materials and systems that operate at the nanoscale and taking advantage of our abilities to visualize, design, and control.

To date, many of the commercial products are based on passive nanostructures incorporated into coatings, polymers, or ceramics where they provide desirable properties to the product. For example, the presence of nanoparticles in a thin film can improve on the permeability properties of the film by simply creating barriers to simple diffusion across the film.

A number of other products take advantage of the unique properties to create active nanomaterials. For example, one of the most commonly used nanomaterials is silver nanoparticles which have been shown to have great anti-bacterial properties, most arising from the enhanced surface reactivity of the small silver nanoparticles. As a result, silver nanoparticles now appear in products as diverse as socks, cutting boards, refrigerators, and washing machines.

The following generation of products will depend on nanosystems that are responsive to their environment to achieve desirable objectives. For example, nanoparticles may be assembled to contain drugs in structures from which they cannot escape unless the particles are subjected to a specific environment which causes a structural change that releases the drug. Such nanosystems are yet to be commercially developed to a great extent, but the research is rapidly moving forward.

Ultimately, the report predicts, nanotechnologies will merge with biotechnologies, information technologies, and cognitive sciences to create large complex, multifunctional technologies that have the power to transform our societies and existence in profound ways — perhaps for the better and perhaps for the worse.

It is important to recognize that in many cases the economic impact may not arise from the nanomaterials themselves, but rather from the products they enable. The value chain will start with the nanomaterials and end with the nano-enabled products as illustrated in Figure 1.1

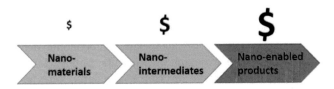

Figure 1.1: Illustration of the value chain of nanotechnology based products.

The nanomaterials will be nanoparticles, nanofibers, nanowires, dendrimers or other similar nano-objects. They may have to be produced in large quantities and therefore will become commodity products where the lowest price is most important. Examples include the use of nanoparticles such as nanoscale zerovalent iron (nZVI) for environmental remediation. Alternatively, they may be produced in small quantities for highly specialized purposes, such as quantum dots that are used in medical diagnostic applications.

The nanointermediates are components that contain nano-objects integrated into a functional unit, such as coatings and printer toners, that can be used as final products, or optical components, memory chips, and sensors that generally are incorporated into the final products. In these intermediates, the nanomaterial may constitute only a very small part, but they are critical for imparting new or better properties. For example, nanoparticles constitute only a few percent by mass in plastic films that are used as permeability barriers, but they improve the permeability properties of the film significantly.

The nano-enabled products are the final products that consumers will purchase and use and include anything from beer bottles to automobiles, from sunscreens to clothing, and from cell phones to robots. The cost of the nanomaterials or nanostructures may only be a small fraction of the total, but the value of the final product depends critically on the performance of the nanomaterials. For example, there are devices that can test for the presence of specific proteins, such as the human chorionic gonadotrophin protein, which is a marker for pregnancy in the early stages. When the protein is present it causes aggregation of nanoparticles coated with antibodies to the protein, which in turn causes a change in color. In the pregnancy tests, the gold nanoparticles act as a specific biosensor, even though they constitute only a fraction of the material or the cost of the test device.

1.4 MOVING FORWARD

The promises of nanoscience and nanotechnology are vast and appear almost unlimited. Just as communications and information technology applications have changed and will continue to change the way we operate as a society in most profound ways — think the internet of things — we may expect nanotechnologies not only to support that ongoing change, but to introduce new paradigms that will have equally surprising and unpredictable impact. To be part of this change in the future, we need be educated and understand better how the nanoscale world works so that we may understand both the positive impact and the potential for negative impact, for there is seldom progress in technology without new ancillary challenges.

There are cautionary tales to learn from the way other technologies emerged in the recent past. For example, biotechnology applications were hampered significantly by public reluctance to accept genetically modified agricultural products. This was in part because some companies were seen to exploit the technology in such a way as to cause exclusive uses and effectively preventing normal agricultural practices. There were therefore significant ethical issues raised in the commercialization processes. Moreover, the ability to sequence entire genomes has created interesting privacy issues — who has the right to know? Likewise, the ability to modify the genome of individuals raises significant questions about whether such technologies should be applied at all.

Correspondingly, in the domain of information technology, the applications have become so pervasive that it is difficult to see how we can live without them. However, the emergence of the cloud and the presence of the internet have exposed weaknesses in terms of security and privacy of information.

It is inevitable that the development of new nanotechnology based applications will raise similar questions of the possible health, environmental, ethical, and legal impact of these applications. For example, one of the keys to the use of nanomaterials is that we can impart new properties to these materials, but that begs the question whether we understand all of the effects of these new properties. Will these materials create unintended consequences for health or environment? These are issues that need to be addressed as the field of nanotechnology matures.

Meanwhile, the excitement is there and I believe that the information in this textbook will allow the reader to prepare for participating in the nanotechnology revolution either as an active creator of new science or new nanotechnologies or as a participant in the incorporation of these into new products.

II

Understanding quantum confinement

Introduction to Part II

Why do quantum dots, nanometer sized particles made of materials such as CdSe, have colors that depend on their sizes? Why are gold nanoparticles red? Why do we expect nanoscale electronic circuits to introduce problematic new effects? There are numerous new questions that confront us when we move to the nanoscale.

Many of the key properties of nanoscale materials derive from the special effects of confining particles, particularly electrons, to small spaces. There are two fundamentally important concepts that emerge: particles confined in space can only exist with specific, discrete energies, which is the concept of quantization of energy; and the description of the particles and their energies depend on the geometry of the confinement, which gives rise to the concept of size and shape dependency of quantum confinement effects.

The first two chapters (Chapters 2 and 3) in this section will define the rules of quantum mechanics while the third chapter (Chapter 4) will provide examples of solutions for a variety of geometries of confinement and finally the fourth chapter (Chapter 5) will examine the consequences for selected applications.

CHAPTER 2

The postulates of quantum mechanics

Much of our understanding of the physical world we live in rests on some key fundamental principles, sometimes we call them laws, from which we can derive expressions for observable phenomena, the observations in turn supporting the validity of our fundamental principles.

For example, Newton's three laws of motion form the foundation for the field of classical mechanics and allow us to explain observations ranging from the movement of a ball after we throw it into the air to the movement of planets in our solar system.

Newton's Laws:

Newton's first law tells us that a body will remain at rest or move at a constant velocity unless there are external forces acting on it.

Newton's second law tells us that if a force is acting on a body, it will accelerate in the direction of the force in proportion to the mass: $\vec{F} = m\vec{a}$.

Newton's third law tells us that when one body exerts a force on another body, the other body exerts a force on the first of equal magnitude but in the opposite direction: $\vec{F_a} = -\vec{F_b}$.

Likewise, as we will discuss in more detail later, the three laws of thermodynamics form the foundation for understanding the flow of energy in physical and chemical systems.

The three laws of thermodynamics:

The First Law tells us that energy is conserved in an isolated system.

The Second Law tells us that the entropy in an isolated system will always increase for a spontaneous process.

The Third Law tells us that it is impossible to attain the temperature of absolute zero degrees Kelvin.

Correspondingly, there is a set of postulates that form the foundation for quantum mechanics. These are statements that control how we approach solving a set of equations so that we may understand the behavior of small particles that do not behave according to Newton's laws.

The field of quantum mechanics was necessitated by the failure of classical mechanics, the laws of Newton, to explain the behavior of atoms and electrons as they were being discovered and studied at the beginning of the 20th century.

There are many versions of the postulates of quantum mechanics. The one used here follows that used by McQuarrie and Simon in their classical physical chemistry text.

2.1   THE SIX POSTULATES OF QUANTUM MECHANICS

Postulate I:
a: Any state[1] of a particle is described by a state function [2] (or a wave function), $\Psi(x, y, z; t)$, which must be well-behaved[3]: i.e., it is continuous,[4] bounded,[5] single-valued,[6] and has continuous partial derivatives, and the following integral exists, is not equal to zero, and has a finite value.

$$\int \Psi^*\Psi d\tau = \int \Psi^*(x, y, z; t)\Psi(x, y, z; t)dxdydz \neq 0. \qquad (2.1)$$

b: The probability that the particle is in the volume element $d\tau = dxdydz$ is given by $\Psi^*\Psi d\tau = |\Psi|^2 d\tau$.

Postulate II:
For every classical state function or physically observable property, M, there is a corresponding linear, quantum mechanical operator, $\hat{M}$.[7]

---

[1]**State:** the property or properties of a system at any given time.
[2]**State (wave) function:** a mathematical description of the property of the system.
[3]**Well-behaved:** a mathematical term to indicate a set of properties including being continuous, bounded, and single-valued.
[4]**Continuous:** there are no breaks in the function.
[5]**Bounded:** the function cannot go to infinity at any point in space.
[6]**Single-valued:** the function can only have one value for any given position or time.
[7]**Operator:** a function or action that changes the function on which it act – it operates on all functions written to the right of the operator.

Postulate III:
If the state of a system (particle) is described by $\Psi(x, y, z; t)$, then the observed value, m, of the classical physical property, M, is given by

$$\widehat{M}\Psi(x, y, z; t) = m\Psi(x, y, z; t). \tag{2.2}$$

This is called the **Eigenvalue Problem** (m is called the **Eigenvalue**[8] and $\Psi(x, y, z; t)$ is the **Eigenfunction**[9]).

Postulate IV:
The **expectation value**, the expected value or the mean value, $\langle m \rangle$, of a physical property, is given by

$$\langle m \rangle = \frac{\int \Psi^* \left[\widehat{M}\Psi\right] d\tau}{\int \Psi^*\Psi d\tau}. \tag{2.3}$$

Postulate V:
The wave function is obtained by solving the **Equation of State**[10] for the Hamiltonian namely the **Time-Dependent Schrödinger Equation**

$$\widehat{H}\Psi(x, y, z; t) = i\hbar\frac{\partial}{\partial t}\Psi(x, y, z; t) = -\frac{\hbar}{i}\frac{\partial}{\partial t}\Psi(x, y, z; t). \tag{2.4}$$

Postulate VI:
The wave function is anti-symmetric for the interchange of any two electrons (or Fermions).

The purpose of these postulates is to provide a framework within which we can, on the one hand determine the wave functions so that we can describe the system of interest at any given time or place, and on the other hand, obtain some physical insights into the systems of interest by relating the mathematics to physical properties familiar within the classical world, the world we can more easily observe and understand.

Postulate I provides the mathematical definition of a function that has no physical meaning in its own right, but is used to describe the state of the particle of interest and, importantly, *the square of its magnitude is related to the probability that we will find the particle somewhere in space at any particular*

---

[8]**Eigenvalue:** the constant value arising from an operator acting on its **Eigenfunction**.(Eigen is German for 'self').
[9]**Eigenfunction:** a function that reproduces itself - except for a constant - after the operator acts on it.
[10]**Equation of state:** an equation that is designed to reveal a property of the system. The Hamiltonian represents the energy of the system.

time. We call it a wave function, since, as we shall see, it has oscillatory properties similar to waves in classical systems such as the waves on the surface of water in a glass that is perturbed.

Postulate II introduces the new concept of an operator, which is, in the simplest form, simply a mathematical object that can change a function in predictable ways. Simple examples of operations are addition, subtraction, multiplication, and division. More complex operations include differentiation, and integration. Postulate II also states that for every physical property that we know from classical mechanics (such as position, momentum, and energy), there is a corresponding operator, which can be used to operate on the wave function to obtain information about that physical property for the particle of interest in our system. Table 2.1 provides a partial list of the physical observable properties and the operators that will be useful for describing them.

Postulate III describes the concept that some operators will operate on a wave function to reproduce the exact same wave function except for a constant number. This constant number represents the value of the physical property corresponding to the operator. For example, if the momentum operator, $\widehat{p}$ operates on a wave function, $\Psi(x, y, z; t)$, to produce the same wave function except for a constant, p, then that constant is the momentum of the particle at that point in space and time:

$$\widehat{p}\Psi(x, y, z; t) = p\Psi(x, y, z; t). \tag{2.5}$$

Since the operation reproduces the same function, the function is called an **eigenfunction** and the constant value is termed the **eigenvalue** and the entire operation is called an **eigenvalue problem**.

Postulate IV recognizes that there will be wave functions that are not eigenfunctions of a particular operator and therefore provides the means whereby we can estimate the mean value — or the expected value — of a particular property of the particle. The notation using the square brackets indicates that the first step is to operate with the operator on the wave function to get the new function, $[M\Psi]$. This new function is then multiplied by the complex conjugate of the wave function, Ψ^*, and the product is integrated over all space and time coordinates. This integral is divided by the integral in Postulate I, which is neither zero nor infinite, to normalize the expectation value. This postulate is critical since it provides the way for us to estimate expected value of any physical property of the system of interest as long as we know what wave functions represent the states of the system even if the wave function is not an eigenfunction of the operator (see later)

Postulate V simply states that, while we could choose to describe a system by any number of wave functions, we choose to select those wave functions [11] that satisfy the **Time-Dependent Schrödinger Equation** (TDSE) so that we can calculate directly the energy of the particle in the system of interest.

---

[11] A set of wave functions that are all solutions to the TDSE is in other contexts termed a set of "basis functions" since they can be used for more advanced calculations using perturbations theory.

Table 2.1: Physical Observables and Corresponding Operators

Observable	Symbol	Symbol	Operation
Observables and Operators			
One-dimensional Systems			
Position	x	$\widehat{x}$	multiply by x
Momentum	p_x	$\widehat{p}_x$	$\frac{\hbar}{i}\left(\frac{\partial}{\partial x}\right)$
Momentum Squared	p_x^2	$\widehat{p}_x^2$	$-\hbar^2\left(\frac{\partial^2}{\partial x^2}\right)$
Kinetic Energy	$K_x = \frac{p_x^2}{2m}$	$\widehat{K}_x = \frac{\widehat{p}_x^2}{2m}$	$-\frac{\hbar^2}{2m}\left(\frac{\partial^2}{\partial x^2}\right)$
Potential Energy	$V(x;t)$	$\widehat{V}(x;t)$	multiply by $V(x;t)$
Total Energy	$E = K_x + V(x;t)$	$\widehat{H} = \widehat{K}_x + \widehat{V}(x;t)$	$-\frac{\hbar^2}{2m}\left(\frac{\partial^2}{\partial x^2}\right) + V(x;t)$
Three-dimensional Systems			
Position	$\boldsymbol{r}$	$\widehat{\boldsymbol{r}}$	multiply by $\boldsymbol{r}$
Momentum	$\boldsymbol{p}$	$\widehat{\boldsymbol{p}}$	$\frac{\hbar}{i}\left(\boldsymbol{i}\frac{\partial}{\partial x} + \boldsymbol{j}\frac{\partial}{\partial y} + \boldsymbol{k}\frac{\partial}{\partial z}\right)$
Momentum Squared	p^2	$\widehat{p}^2$	$-\hbar^2\left(\frac{\partial^2}{\partial x^2} + \frac{\partial^2}{\partial y^2} + \frac{\partial^2}{\partial z^2}\right)$
Kinetic Energy	$K = \frac{p^2}{2m}$	$\widehat{K} = \frac{\widehat{p}^2}{2m}$	$-\frac{\hbar^2}{2m}\left(\frac{\partial^2}{\partial x^2} + \frac{\partial^2}{\partial y^2} + \frac{\partial^2}{\partial z^2}\right)$
Potential Energy	$V(\boldsymbol{r};t)$	$\widehat{V}(\boldsymbol{r};t)$	multiply by $V(\boldsymbol{r};t)$
Total Energy	$E = K + V(\boldsymbol{r};t)$	$\widehat{H} = \widehat{K} + \widehat{V}(\boldsymbol{r};t)$	$-\frac{\hbar^2}{2m}\widehat{\nabla}^2 + \widehat{V}(\boldsymbol{r};t)$
Angular Momentum	$l_x = yp_z - zp_y$	$\widehat{\boldsymbol{L}}_x$	$-i\hbar\left(y\frac{\partial}{\partial z} - z\frac{\partial}{\partial y}\right)$
	$l_y = zp_x - zp_z$	$\widehat{\boldsymbol{L}}_y$	$-i\hbar\left(z\frac{\partial}{\partial x} - x\frac{\partial}{\partial z}\right)$
	$l_z = xp_y - yp_x$	$\widehat{\boldsymbol{L}}_z$	$-i\hbar\left(x\frac{\partial}{\partial y} - y\frac{\partial}{\partial x}\right)$

Postulate VI recognizes that in most cases we are investigating electrons in a system, and since electrons are particles with spin $\frac{1}{2}$ — also called Fermions — there is an additional constraint on the wave function, which allows two electrons of opposite spin to occupy the same energy state.

Having established these postulates, the process for proceeding is in principle straightforward:

- First, establish what the Hamiltonian operator is for the system of interest; in the simplest cases this involves determining what the potential energy operator is since the kinetic energy operator is the same for all of them.

- Second, solve the Time-Dependent Schrödinger Equation as defined in Postulate V to get the wave functions of the system.

- Third, use the wave functions to calculate the expectation values of the physical properties of interest — energy, momentum, position, etc. — by using Postulate IV.

In this and the following chapters of this section we will consider first several simple systems that illustrate some fundamental principles and then some more complex systems that are better representations of real systems.

It should be noted that there is an alternative approach to quantum mechanics developed by Heisenberg based on matrix mechanics as compared to the approach used here which is based on wave mechanics. The two approaches lead to the same overall conclusions. Historically, the Schrödinger wave mechanics approach is favored by chemists and the Heisenberg matrix mechanics approach is favored by physicists.

2.2 COMMUTATORS AND THE PRINCIPLE OF SHARED EIGEN-FUNCTIONS

Consider two operators $\hat{A}$ and $\hat{B}$. In general, we expect that operating first with operator $\hat{B}$ and then with operator $\hat{A}$ may give a different result than operating first with operator $\hat{A}$ and then with operator $\hat{B}$. To test this, we define the **commutator** of these operators as $\left[\hat{A}, \hat{B}\right] = \hat{A}\hat{B} - \hat{B}\hat{A}$ and we say that the operators commute if the order of operation does not matter and hence the commutator is equal to zero.

Consider that each operator satisfies an eigenvalue problem with corresponding eigenfunctions Ψ_A and Ψ_B and eigenvalues a and b, then if $\hat{A}$ and $\hat{B}$ commute we can show that they can simultaneously have the same eigenfunctions:

$$\hat{A}\left(\hat{B}\Psi_A\right) = \hat{B}\left(\hat{A}\Psi_A\right) = \hat{B}\left(a\Psi_A\right) = a\left(\hat{B}\Psi_A\right). \tag{2.6}$$

The first and the last term show that if the operators commute, then the function arising from the operation of $\hat{B}$ on Ψ_A, namely $\left(\hat{B}\Psi_A\right)$, is also an eigenfunction of $\hat{A}$. Whenever the eigenvalues are unique to particular eigenfunctions, then it follows that $\hat{B}\Psi_A = (constant)\Psi_A$, or that Ψ_A is also an eigenfunction of the operator $\hat{B}$. Correspondingly, it follows that $\hat{A}\Psi_B = (constant)\Psi_B$ and that Ψ_B is also an eigenfunction of the operator $\hat{A}$. In other words, if the operators commute and the commutator is zero, then the two operators share the same set of eigenfunctions. The implication of Postulate III is that the physical observables of commuting operators can be determined simultaneously. Conversely, the physical observables of operators that do not commute cannot be determined simultaneously. This gives rise to the **uncertainty principle**.

CHAPTER 3

Time-dependent and time-independent Schrödinger equations

3.1 CONSERVATIVE FORCES

The Time-Dependent Schrödinger Equation (TDSE) is provided in Postulate V as the means whereby the wave functions can be determined. This requires that the Hamiltonian Operator can be defined. In general, the Hamiltonian can be a complex function of position and time (some examples will be provided later), but we will proceed with some constraints on the potential energy term, which simplifies the problem substantially.

Specifically, we will consider first the Hamiltonian being exclusively a sum of kinetic energy of a particle and the potential energy in which it moves so that, from Table 2.1 we get

$$\widehat{H}\Psi(x,y,z;t) = \left[-\frac{\hbar^2}{2m}\widehat{\nabla}^2 + \widehat{V}(r;t) \right] \Psi(x,y,z;t) = \frac{\hbar}{i}\frac{\partial}{\partial t}\Psi(x,y,z;t). \quad (3.1)$$

If, in the first instance, we restrict our attention to systems exposed to conservative forces only, which correspond to potential energies that are independent of time ($\widehat{F} = -\nabla\widehat{V}(r)$), then there is no time dependence in the Hamiltonian Operator. The left-hand side of the differential equation is then only dependent on position while the right-hand side of the equation is only dependent on time. The equality is only possible if both sides are equal to the same constant and that permits a separation of variables and determination of the time-dependence of the wave function by solving the first order differential equation in time

$$-\frac{\hbar}{i}\frac{\partial}{\partial t}f(t) = E. \quad (3.2)$$

This has the simple solution that $f(t) = e^{-iEt/\hbar} = e^{-i\omega t}$, where we have used $E = \hbar\omega = h\nu$. [The details of the separation of variables and the solution leading to the constant representing energy is left to an advanced problem assignment.]

One of the consequences of restricting our attention to conservative forces is that we can rewrite the wave function as the product of a time-independent wave function, $\psi(x, y, z)$, and the time-dependent component of the wave function, $e^{-i\omega t}$, as follows

$$\Psi(x, y, z; t) = \psi(x, y, z)e^{-i\omega t}. \tag{3.3}$$

Even though these wave functions are dependent on time, they are termed stationary state wave functions because the physical properties derived from them are independent of time since calculating them will always involve an integral with a time-independent operator so that

$$\int \Psi^* \left[\widehat{M}\Psi\right] = \int \psi^*(x, y, z)e^{i\omega t}) \left[\widehat{M}(x, y, z)\psi(x, y, z)e^{-i\omega t}\right] d\tau$$

$$= \int \psi^*(x, y, z) \left[\widehat{M}(x, y, z)\psi(x, y, z)\right] d\tau. \tag{3.4}$$

Substituting the new form of the wave function into the TDSE yields the Time-Independent Schrodinger Equation (TISE)

$$\widehat{H}\psi(x, y, z) = E\psi(x, y, z). \tag{3.5}$$

This shows that the time-independent wave function is an eigenfunction of the Hamiltonian Operator and that the eigenvalue is the energy. The significance of the TISE is that if we can solve the differential equation, we have determined the energies of the states of the systems that are described by the set of stationary state wave functions.

CHAPTER 4

Solving the time-independent Schrödinger equation

The solutions to the Time-Independent Schrödinger Equation (TISE) can be found as long as we can define the Hamiltonian, which in the simplest cases, means defining the potential in which the particle of interest is moving. In the following sections, we will consider a single particle of mass, m_p, moving in a potential that is time-independent, but can vary in space. We will follow four steps:

- Define the Hamiltonian by defining the Potential Energy Operator since $\widehat{H} = -\frac{\hbar^2}{2m_p}\widehat{\nabla}^2 + \widehat{V}(\boldsymbol{r})$.

- Solve the TISE to get the wave functions $\psi(x, y, z)$ and then the total stationary state wave functions $\Psi(x, y, z; t) = \psi(x, y, z)e^{-i\omega t}$.

- Determine the energies of the various states of the system.

- Examine the implications of the solutions to the Schrödinger equations.

We will progress from the simplest systems to more complex systems, starting with one-dimensional problems and moving towards three-dimensional systems of various geometries. As we progress, the mathematical treatment will become less rigorous, but we will still be able to examine the key trends of introducing confinement on the movement of the particle.

4.1   THE FREE PARTICLE

Let us consider first a particle of mass, m_p, moving in space where the potential energy is zero everywhere, which means that $V(\boldsymbol{r}) = 0$ for all values of $\boldsymbol{r}$.

Further, let us initially chose a coordinate system in which the particle moves along the x-axis so that it becomes a 1-dimensional problem. Then the TISE becomes

$$-\frac{\hbar^2}{2m_p}\widehat{\nabla}^2\psi(x) = -\frac{\hbar^2}{2m_p}\frac{\partial^2}{\partial^2 x}\psi(x) = E\psi(x). \tag{4.1}$$

This can be re-written as

$$\frac{\partial^2}{\partial^2 x}\psi(x) + \frac{2m_p E}{\hbar^2}\psi(x) = \frac{\partial^2}{\partial^2 x}\psi(x) + k^2\psi(x) = 0. \tag{4.2}$$

This differential equation has a general solution of the form

$$\psi(x) = \psi^+(x) + \psi^-(x) = Ae^{ikx} + Be^{-ikx}$$
$$= (A+B)cos(kx) + i(A-B)sin(kx). \tag{4.3}$$

Here, $k = \frac{\sqrt{2m_p E}}{\hbar}$ and the latter part is obtained by application of the Euler Formula.

Euler Formula: $e^{ix} = cosx - isinx$

The solution will hold for all values of energy $E \geq 0$, but if $E < 0$, then ik is real and the wave functions are not bounded and hence not valid by Postulate Ia.

The momentum operator from Table 2.1 is $\frac{\hbar}{i}\frac{\partial}{\partial x}$ and it follows that the general wave function is not an eigenfunction of the momentum operator and the momentum is then not well defined. However, the individual parts of the wave functions are eigenfunctions of the momentum operator with defined momentum, for example:

$$\widehat{p}\psi^+(x) = \frac{\hbar}{i}\frac{\partial}{\partial x}e^{ikx} = \frac{\hbar}{i}ikAe^{ikx} = \hbar k\psi^+(x). \tag{4.4}$$

This shows that the momentum $p = \hbar k = \sqrt{2m_p E}$ is positive and real, suggesting that this part of the wave function, $\psi^+(x)$, describes a particle moving in the direction of positive values of x. Correspondingly, $\psi^-(x)$ describes a particle moving in the direction of negative values of x with a momentum $p = -\hbar k = -\sqrt{2m_p E}$.

Postulate Ib allows us to calculate the probability of finding the particle in the small section of space, dx by $\psi^*(x)\psi(x)dx = A^*e^{-ikx}Ae^{ikx} = |A|^2$, which is a constant for all values of x. There is therefore a finite and constant probability of finding the particle everywhere.

While the momentum is defined, the position is completely unknown.

Note that if the particle is moving in a constant potential $V_0 \geq 0$, the solution is fundamentally the same, but the momentum, and hence the wavelength, will be reduced (see problem assignment). We will examine this further in Section 4.4 where we consider a particle approaching a potential barrier.

4.1.1 The uncertainty principle at work

Table 2.1 defines the momentum and position operators as $\frac{\hbar}{i}\frac{\partial}{\partial x}$ and x respectively. It is easy to show that the momentum and position operators do not commute and that the wave function for the free particle is not an eigenfunction of either the momentum operator or of the position operator (see problems). This means that neither the momentum nor the position of the particle can be known with certainty at the same time.

Postulate IV tells us how to calculate the expected values of the momentum and the position as

$$\langle p \rangle = \frac{\int \psi^*(x)\left[\widehat{p}\psi(x)\right]dx}{\int \psi^*(x)\psi(x)dx} \tag{4.5}$$

$$\langle x \rangle = \frac{\int \psi^*(x)\left[\widehat{x}\psi(x)\right]dx}{\int \psi^*(x)\psi(x)dx}. \tag{4.6}$$

Likewise one can calculate the expected value of the variance of each parameter, for example the variance of the momentum is $\langle \Delta p^2 \rangle = \left\langle (\widehat{p} - \langle p \rangle)^2 \right\rangle$ or

$$\langle \Delta p^2 \rangle = \frac{\int \psi^*(x)\left[(\widehat{p} - \langle p \rangle)^2 \, \psi(x)\right]dx}{\int \psi^*(x)\psi(x)dx}. \tag{4.7}$$

It is possible to show (see for example Herbert L. Strauss *Quantum Mechanics: An Introduction*) that

$$\langle \Delta p \rangle^2 \langle \Delta x \rangle^2 \geq -\frac{1}{4}\left[i\hbar\right]^2 \tag{4.8}$$

or that

$$\langle \Delta p \rangle \, \langle \Delta x \rangle \geq \frac{1}{2}\hbar. \tag{4.9}$$

This tells us that, in general, the product of the uncertainties in momentum and position is always greater than a finite quantity and can never be zero. This is one manifestation of the Heisenberg Uncertainty Principle that tells us that if the momentum is known with great precision, the position is very uncertain and vice versa, if the position is known with great precision, then the momentum is very uncertain.

The general formulation of the Heisenberg Uncertainty principle for two operators $\widehat{A}$ and $\widehat{B}$ is

$$\sigma_A^2 \sigma_B^2 \geq -\frac{1}{4}\left[\int \phi^* \left[\widehat{A}\widehat{B}\right]\phi d\tau\right]^2$$

. This shows that if the operators commute then the product of the variances σ^2 or uncertainties can approach zero, but operators that do not commute cannot be determined with precision simultaneously (Strauss).

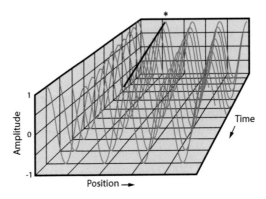

Figure 4.1: Propagation of a wave as a function of position and time. The position of the peak moves as time evolves.

4.1.2   The concept of wave packets

In the previous section, we determined the wave functions using the TISE. The corresponding total wave functions are obtained by multiplying with the time-dependent components

$$\Psi(x, y, z; t) = \psi(x, y, z)e^{-i\omega t} = \left(Ae^{ikx} + Be^{-ikx}\right)e^{-i\omega t}$$
$$= \left(Ae^{i(kx-\omega t)} + Be^{i(-kx-\omega t)}\right) \qquad (4.10)$$

The only difference between the two terms in the last equation is the sign of the parameter, k, which suggests that the first term corresponds to a particle with a positive momentum and the second term to that with a negative momentum.

Each term represents a standing wave moving in either the positive or in the negative direction with a spatial wavelength related to the deBroglie wavelength $\lambda = \frac{h}{p} = \frac{2\pi}{k}$ and a temporal period of $T = \frac{1}{\nu} = \frac{2\pi}{\omega}$. For a particle with a defined momentum or defined energy moving along the positive direction of x

$$\Psi(x; t) = Ae^{ikx}e^{-i\omega t} = A'\cos(kx - \omega t). \qquad (4.11)$$

The propagation of this wave is shown as a function of both position and time in Figure 4.1.

Figure 4.2 shows the propagation for a wave at twice the spatial wavelength.

The phase velocity is defined as the distance a particular phase of the wave, such as the peak, moves as a function of time. It can be seen to be the wavelength (distance) moved in one period (time) and as a consequence depends only on their ratio, i.e., $v_p = \frac{\lambda}{T} = \frac{\omega}{k}$. The phase velocity is therefore independent of k and ω individually. This is evident when comparing the

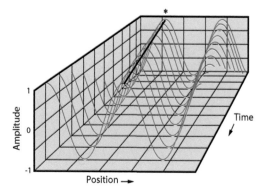

Figure 4.2: Propagation of a wave at twice the wavelength shows that the peak moves at the same rate as indicated by the solid black line.

propagating waves in Figures 4.1 and 4.2 where the peak moves the same distance along the position axis for a given time along the time axis. This is best visualized by following the peaks marked by the * symbol and indicated by the black line.

The propagating waves illustrated in Figures 4.1 and 4.2 describe particles with a precisely known k-value, that is a precisely known momentum or energy of the particle. If there is an uncertainty in the energy, there will be an uncertainty in the values of k and hence in the wavelengths $[E = \frac{\hbar^2 k^2}{2m_p}]$. The particle will then be described by a superposition of wave functions that depend on the range of k-values as defined by the function $f(k)$, such that

$$\Psi(x; t) = \sum f(k) \cos(kx - \omega t). \qquad (4.12)$$

Assuming that there is one point in space and time when the waves are in phase, then these waves will interfere constructively around that point. This effect is shown in Figure 4.3 where a series of waves of different frequencies and of amplitudes given by $f(k)$, all in phase at the center of the position scale, are superimposed. The result is a damped oscillation of limited dimension along the x-axis and decaying in both directions from the point where they are all in phase as shown in Figures 4.3 and 4.4. This superposition of waves defines a wave packet.

Since the phase velocity is independent of the wavelength, the point at which the waves are in phase will move at the same velocity for all the superimposed waves and hence the wave packet will move along the x-axis at a constant phase velocity as long as there is no change in refractive index. The actual velocity of the wave packet is given by the group velocity which is

$$v_g = \frac{\lambda}{T} = \frac{d\omega}{dk} = \frac{c}{n} - \frac{ck}{n^2}\frac{dn}{dk}, \qquad (4.13)$$

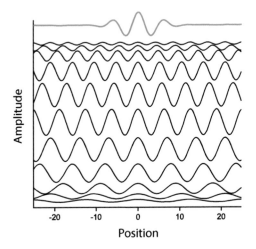

Figure 4.3: Superposition of waves form a wave packet by interference at increasing distances from the center.

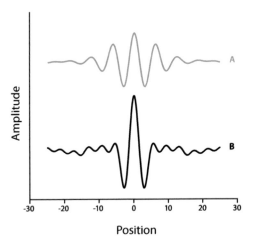

Figure 4.4: Wave packets for a smaller Δk in A and a larger Δk in B.

which is the same as the phase velocity in the limit when $\frac{dn}{dk} = 0$.

The refractive index is related to the phase velocity by the speed of light:
$$n = \frac{c}{v_p} = \frac{ck}{\omega}$$

The wave packet provides a further illustration of the uncertainty principle: when there is uncertainty in momentum, the position gets better defined; the greater the uncertainty in momentum is, the more waves are superimposed and the narrower the distribution of the wave packet along the x-axis becomes, as illustrated in Figure 4.4 by comparing the wave packets in A and B.

The concept of the wave packet is important for understanding the wave-particle duality of photons, electrons, and all other systems described by a wave function. The concept extends to superposition of mechanical vibrations in lattices leading to phonons, to movement of free electrons in metals leading to plasmons, and to many other phenomena collectively known as pseudo-particles. We shall return repeatedly to the concept of wave packets.

4.2 PARTICLES CONSTRAINED BY INFINITE POTENTIALS

The previous section analyzed the consequences of solving the Schrödinger Equation for a particle moving freely in space in a constant potential $V = 0$ (and in the problem assignment when $V > 0$). This is useful for understanding the concepts of waves and the formation of wave packets. However, there are no restrictions on the energy available to the particle since it is free to move everywhere.

In order to understand the consequences of quantum confinement, we need to consider when the particle is constrained in space by potentials that are greater than the energy available to the particle. In this section, we will consider particles constrained by infinitely high potentials, that is, $V \to \infty$ everywhere outside the region of interest. We will progress from a one-dimensional systems (1-D box), to two-dimensional rectangular or cylindrical boxes, to three-dimensional boxes and spheres, thereby approaching more and more realistic systems.

4.2.1 One-dimensional box bounded by infinite potentials

The one-dimensional box with infinite walls is illustrated schematically in Figure 4.5 in which we consider a particle of mass, m_p, constrained to a line on the x-axis between the positions $x = 0$ and $x = b$, where b is the length of the box.

Classically, this would be like a pearl on a string confined by two knots. We would expect the pearl to be located at a specific position somewhere on the string at any given time. However, to determine the properties of the quantum particle, we need to solve the Schrödinger Equation for all values of x. We will start with the TISE in one dimension:

$$\widehat{H}\psi(x) = E\psi(x). \tag{4.14}$$

Outside the box, when $x < 0$ or $x > b$, the particle would be subjected to an infinite potential, which suggests that the particle cannot exist in these

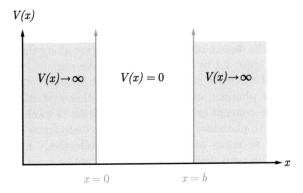

Figure 4.5: The one-dimensional box with infinite walls.

regions. We restate this as the probability of finding the particle in these regions being identically equal to zero. By Postulate Ib

$$|\psi(x)|^2 = 0 \Rightarrow \psi(x) = 0 \tag{4.15}$$

for all $x < 0$ and all $x > b$ Thus, the wave function is zero outside the box.

Inside the box, when $0 < x < b$, the potential is zero, so the TISE becomes

$$-\frac{\hbar^2}{2m_p}\frac{\partial^2}{\partial x^2}\psi(x) = E\psi(x). \tag{4.16}$$

This must have the same general solution as in the case of the free particle and hence

$$\psi(x) = \psi^+(x) + \psi^-(x) = Ae^{ikx} + Be^{-ikx}$$
$$= (A + B)cos(kx) + i(A - B)sin(kx) \tag{4.17}$$

for $0 < x < b$. This suggests that the particle is described by some type of wave inside the box.

The alert reader will have noted that so far we have not considered the cases where $x = 0$ or $x = b$, in other words, we have not considered what happens at the edges, which provide constraints that we call boundary conditions.

At the edges of the box, Postulate Ia demands that the wave functions must be continuous, which means that the limits of the wave function as it approaches both edges from both the positive and the negative direction must be the same:

$$lim_{x \to 0^+}\psi(x) = lim_{x \to 0^-}\psi(x) = 0 \tag{4.18}$$
$$lim_{x \to b^+}\psi(x) = lim_{x \to b^-}\psi(x) = 0. \tag{4.19}$$

In other words, the wave function is zero at the edges. These boundary conditions allow us to solve for the constants A and B in the general solutions to the TISE within the box.

Substituting for $x = 0$ yields

$$\psi(0) = Ae^{ik0} + Be^{-ik0} = A + B = 0 \text{ so } A = -B. \tag{4.20}$$

Thus

$$\psi(x) = A\left(e^{ikx} - e^{-ikx}\right) = 2iA\sin(kx). \tag{4.21}$$

Further substituting $x = b$ yields

$$\psi(b) = 2iA\sin(kb) = 0. \tag{4.22}$$

The sine function is zero whenever the argument is an integer multiple of π, which requires that $kb = n\pi$ for $n = 0, 1, 2, 3, \ldots$ [Note: $n = 0$ means $\psi_n(x) = 0$ for all x which is a trivial solution since there would be no particle anywhere and hence we require $n > 0$]

We can normalize the wave function by setting the integral of the probability distribution (Postulate I) equal to unity (see problem assignment)

$$\int \psi^*(x)\psi(x)dx = 1. \tag{4.23}$$

This yields a value for the constant $A = \frac{1}{i\sqrt{2b}}$ and therefore the final form of the wave functions become

$$\psi_n(x) = 2iA\sin\left(\frac{n\pi}{b}x\right) = \sqrt{\frac{2}{b}}\sin\left(\frac{n\pi}{b}x\right) \tag{4.24}$$

with $n = 1, 2, 3\ldots$ and $0 \le x \le b$.

It is evident that there will be an infinite number of wave functions, each associated with an integer n, which is the quantum number for the nth state of the systems that is characterized by the wave function, $\psi_n(x)$.

We can also calculate the expected energy associated with each of the states using either Postulate III or Postulate IV (see problems assignment) to get

$$E_n = \frac{n^2 h^2}{8m_p b^2}. \tag{4.25}$$

There are a number of implications of these results:

1. The wave functions correspond to an infinite set of standing waves of different wavelength and energies that depend only on the quantum number n, which is an integer greater than zero.

2. The wavelength of the standing wave in the nth state is $\lambda_n = \frac{h}{|p|} = \frac{h}{k\hbar} = \frac{h}{nh/2b} = \frac{2b}{n}$. As the quantum number increases the wavelength decreases linearly.

3. The probability of finding the particle at a given region between x and $x+$ dx in the box will be periodic and given by $|\psi(x)|^2 \, dx = \frac{2}{b} sin^2 \left(\frac{n\pi}{b} x \right) dx$.

4. For all of the states with $n > 1$, there will be positions in the box, other than at the edges, where the probability of finding the particle will be identically equal to zero. These are called nodes and there are $n - 1$ nodes in the nth state of the system.

5. There are only certain energies that are allowed; the energy is quantized and depends on the square of the quantum number and the difference in energy between adjacent levels increases linearly with the quantum number since $E_{n+1} - E_n = \Delta E_n = (2n + 1) \frac{h^2}{8m_p b^2}$. This concept will be important later when we discuss the energy needed to effect transitions of the particle between energy levels (see problem assignment).

6. The lowest possible energy available is greater than zero since $E_1 = \frac{h^2}{8m_p b^2}$; there is a zero-point energy.

7. The energy depends inversely on the mass of the particle. Lighter particles have larger energies for a given length of the box. Conversely, as the mass, m_p, becomes very large, the energy levels become very close and in the limit will approach the classical particle that can be at any energy level.

8. The energy depends inversely on the square of the length of the box, b. This is the first example of many in which the size of the system affects the energy of the system. This is the root of the quantum confinement effect in nanoscale systems. As the size of the box increases, the energy level separation decreases and in the limit of a very large box, the quantum levels merge and the particle approaches a free particle that can assume any energy greater than zero

These effects are illustrated in Figures 4.6 A and B which show respectively the wave functions and the probability distributions for a single particle in a 1-D box with infinite walls. These functions are shown for the first six quantum numbers ($n = 1, 2, 3, 4, 5,$, and 6) offset at the corresponding energy levels and normalized in their amplitude.

The oscillatory nature of the standing wave of the time-independent wave function is evident for each energy level and it is clear that the wavelength decreases as the quantum number increases.

The probability distribution is also seen to be periodic, with a number of nodes where the probability is zero. The number of nodes increases linearly with the quantum number.

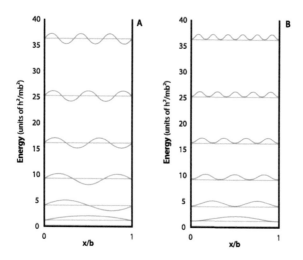

Figure 4.6: A: The wave functions for the 1-D box with infinite walls. B: The probability distributions for the particle in the 1-D box.

The web-site: http://www.falstad.com/qm1d/ contains an 1-D Quantum States Applet that runs in Java which allows you to observe the wave functions, the probability distributions and the momentum of the particle in a box [as well as a number of more advanced properties]. This Applet also allows you to adjust the parameters of the particle in a box, such as the mass of the particle and length of the box. This same Applet allows you to examine other 1-dimensional systems that we will discuss later in this section. The web-site is an excellent opportunity to explore in depth the properties of 1-D quantum systems.

We will explore some of the direct applications of this relatively simple system later in this section, but for the present, consider that it describes the confinement in one direction only and that it can therefore be applied to thin films of material, in which particles such as electrons can flow freely in the plane of the film (using the formulation of the wave function for a free particle in this plane) but are confined in the direction perpendicular to the film. This leads to quantum confinement along that direction as described by this particle in the box analysis and the energies will depend on the film thickness. In the world of nanoscience and nanotechnology, these are nanoplates.

4.2.2 Two- and three-dimensional boxes bounded by infinite potentials

The logic used to solve the 1-dimensional particle-in-a-box problem can be extended to two- and three-dimensional boxes by employing the principle that the coordinates are independent.

Consider first the problem illustrated in Figure 4.7 for a particle of mass

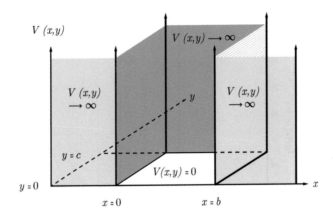

Figure 4.7: The two-dimensional box with infinite walls.

m_p in a two-dimensional box. The Time-Independent Schrödinger Equation (TISE) for this problem is

$$-\frac{\hbar^2}{2m_p}\left(\frac{\partial^2}{\partial x^2} + \frac{\partial^2}{\partial y^2}\right)\psi(x,y) = E\psi(x,y). \qquad (4.26)$$

Since x, and y are independent coordinates, the wavefunction can be expressed as a product of two wavefunctions, $\phi(x)$ and $\phi(y)$, which are functions of only x and only y, respectively:

$$\psi(x,y) = \phi(x)\phi(y). \qquad (4.27)$$

The TISE can then be solved for each co-ordinate separately, that is, the solution is that of the 1-dimensional particle in the box in each coordinate, and we get

$$\phi_{n_x}(x) = \sqrt{\frac{2}{b}}sin\left(\frac{n_x\pi}{b}x\right) \qquad (4.28)$$

$$\phi_{n_y}(y) = \sqrt{\frac{2}{c}}sin\left(\frac{n_y\pi}{c}y\right). \qquad (4.29)$$

The total, time-independent wavefunction now is

$$\psi_{n_x n_y}(x,y) = \sqrt{\frac{2}{b}}sin\left(\frac{n_x\pi}{b}x\right)\sqrt{\frac{2}{c}}sin\left(\frac{n_y\pi}{c}y\right). \qquad (4.30)$$

We note that the wave function now depends on two quantum numbers — n_x and n_y — one for each coordinate. Each quantum number can take on

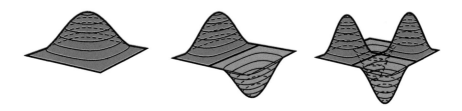

Figure 4.8: The three lowest energy wave functions for the two-dimensional box with infinite walls.

integer values greater than zero. The wave functions also depend on the length of the box in each direction, b and c, respectively, so the standing wave may have different wavelengths in each of the two directions. Solving for the energy, we find that the energies will add

$$E_{n_x n_y} = E_{n_x} + E_{n_y} = \frac{n_x^2 h^2}{8 m_p b^2} + \frac{n_y^2 h^2}{8 m_p c^2}. \tag{4.31}$$

with $n_x = 1, 2, 3, \ldots$ and $n_y = 1, 2, 3, \ldots$

It is illustrative to consider the special case of a square box where $b = c$ for which the wave functions and energies become

$$\psi_{n_x n_y}(x, y) = \frac{2}{b} sin\left(\frac{n_x \pi}{b} x\right) sin\left(\frac{n_y \pi}{b} y\right). \tag{4.32}$$

$$E_{n_x n_y} = E_{n_x} + E_{n_y} = \frac{\left(n_x^2 + n_y^2\right) h^2}{8 m_p b^2}. \tag{4.33}$$

with $n_x = 1, 2, 3, \ldots$ and $n_y = 1, 2, 3, \ldots$

Three of the lowest wave functions, ψ_{11}, ψ_{21}, and ψ_{22} for the two-dimensional square box are illustrated in Figure 4.8. This result allows us to introduce the concept of degeneracy, which refers to the situation where two states of the system, described by different wave functions, will have the same energy. Compare as an example the wave function identified by the quantum numbers $n_x = 1$ and $n_y = 2$ with that identified by the quantum numbers $n_x = 2$ and $n_y = 1$. These wave functions are shown in Figure 4.9 and they differ since the wavelength along the x-axis is different from that along the y-axis. However, both have the energy

$$E_{1,2} = \frac{(1 + 4) h^2}{8 m_p b^2} = E_{2,1} = \frac{(4 + 1) h^2}{8 m_p b^2}. \tag{4.34}$$

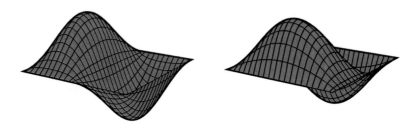

Figure 4.9: Illustration of two degenerate wave functions for $n_x = 1$ and $n_y = 2$ (left) $n_x = 2$ and $n_y = 1$ (right).

There are clearly a number of degenerate states whenever the quantum numbers can be exchanged (see problem assignments).

The nature of these wave functions and the corresponding energies can be explored further using the web-site http://www.falstad.com/qm2dbox/ where it is possible to explore and visualize wave functions for a number of combinations of quantum numbers.

Extension to the three-dimensional box bounded by infinite potentials is in principle trivial since all three coordinates are independent. Thus the wave functions and energies are given by

$$\psi_{n_x n_y n_z}(x, y, z) = \sqrt{\frac{2}{b}} sin\left(\frac{n_x \pi}{b} x\right) \sqrt{\frac{2}{c}} sin\left(\frac{n_y \pi}{c} y\right) \sqrt{\frac{2}{d}} sin\left(\frac{n_z \pi}{d} z\right), \quad (4.35)$$

$$E_{n_x n_y n_z} = E_{n_x} + E_{n_y} + E_{n_z} = \frac{n_x^2 h^2}{8 m_p b^2} + \frac{n_y^2 h^2}{8 m_p c^2} + \frac{n_z^2 h^2}{8 m_p d^2}. \quad (4.36)$$

The case of a cubic box and the consequences for degeneracies will be explored further in the problem assignments.

The result of the particle in a two-dimensional box system can in principle be applied to wires, and therefore to the behavior of electrons in a nanowire in which the electron is confined to the x and y directions but can move freely along the z-direction.

Correspondingly, the result of the particle in a three-dimensional box can in principle be applied to nanoparticles, particularly nanocubes, in which the electrons are confined in all three directions.

We will explore these applications further once we have considered systems confined by finite potentials.

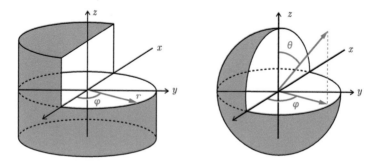

Figure 4.10: Depiction of the cylindrical and spherical coordinate systems.

4.2.3 Rings, disks, shells, and spheres bounded by infinite potentials

In the previous sections we considered particles confined to one-, two- and three-dimensional rectilinear systems that are best considered in the Cartesian coordinate system (x,y,z) allowing us to consider possible applications to thin films or nanoplates, rectangular nanowires, and cuboidal nanoparticles. In all cases, we observe wave functions that are sinusoidal with wavelengths that depend on the size of the systems and we establish that energy is quantized and the energy levels depend inversely on both the mass of the particle and the square of the dimension of the system.

In this section we will extend the concept of quantum confinement to systems that are best considered in cylindrical (r,ϕ,z) or spherical (r,θ,ϕ) coordinate systems (Figure 4.10). This will lead to possible applications to rings, disks, shells and spheres, and eventually to understanding concepts of angular momentum, spherical harmonics, the rigid rotator, and the Hydrogen Atom. To address these problems, we must first express the Schrödinger Equations in these coordinate systems. This is achieved by a transformation of coordinates. For cylindrical coordinates, this means that $x = r\cos\phi$, $y = r\sin\phi$, and $z = z$, whereas for spherical coordinates, this means that $x = r\sin\theta\cos\phi$, $y = r\sin\theta\sin\phi$ and $z = r\cos\theta$. This in turn (see problem assignments) leads to new expressions for the partial derivatives and the Laplacian for cylindrical coordinates:

$$\frac{\partial}{\partial x} = \cos\phi \frac{\partial}{\partial r} - \frac{\sin\phi}{r}\frac{\partial}{\partial \phi} \tag{4.37}$$

$$\frac{\partial}{\partial y} = \sin\phi \frac{\partial}{\partial r} + \frac{\cos\phi}{r}\frac{\partial}{\partial \phi} \tag{4.38}$$

$$\frac{\partial}{\partial z} = \frac{\partial}{\partial z} \tag{4.39}$$

and hence

$$\widehat{\nabla}^2 = \left(\frac{\partial^2}{\partial x^2} + \frac{\partial^2}{\partial y^2} + \frac{\partial^2}{\partial z^2}\right) = \left(\frac{\partial^2}{\partial r^2} + \frac{1}{r}\frac{\partial}{\partial r} + \frac{1}{r^2}\frac{\partial^2}{\partial \phi^2} + \frac{\partial^2}{\partial z^2}\right). \tag{4.40}$$

Similarly for spherical coordinates:

$$\frac{\partial}{\partial x} = \sin\theta\cos\phi\frac{\partial}{\partial r} + \frac{\cos\theta\cos\phi}{r}\frac{\partial}{\partial \theta} - \frac{\sin\phi}{r\sin\theta}\frac{\partial}{\partial \phi} \tag{4.41}$$

$$\frac{\partial}{\partial y} = \sin\theta\sin\phi\frac{\partial}{\partial r} + \frac{\cos\theta\sin\phi}{r}\frac{\partial}{\partial \theta} + \frac{\cos\phi}{r\sin\theta}\frac{\partial}{\partial \phi} \tag{4.42}$$

$$\frac{\partial}{\partial z} = \cos\theta\frac{\partial}{\partial r} - \frac{\sin\theta}{r}\frac{\partial}{\partial \theta} \tag{4.43}$$

and hence

$$\widehat{\nabla}^2 = \left(\frac{\partial^2}{\partial x^2} + \frac{\partial^2}{\partial y^2} + \frac{\partial^2}{\partial z^2}\right)$$

$$= \left(\frac{1}{r^2}\frac{\partial}{\partial r}\left(r^2\frac{\partial}{\partial r}\right) + \frac{1}{r^2\sin\theta}\frac{\partial}{\partial \theta}\left(\sin\theta\frac{\partial}{\partial \theta}\right) + \frac{1}{r^2\sin^2\theta}\frac{\partial^2}{\partial \phi^2}\right). \tag{4.44}$$

These expressions for the Laplacian can be substituted into the Hamiltonian in the Schrödinger Equations to be solved for the appropriate potential energies and geometries.

4.2.3.1 Confinement to a ring bounded by an infinite potential

The case of a particle of mass m_p moving on a ring of radius r_0 in the x, y-plane with a potential of zero on the ring and an infinite potential elsewhere is best treated in cylindrical coordinates with $r = r_0$ being a constant and $z = 0$ as well, so that the derivatives with respect to the r- and z-coordinate disappear. This gives a time-independent form of the Schrödinger equation as follows:

$$-\frac{\hbar^2}{2m_p}\left(\frac{\partial^2}{\partial r^2} + \frac{1}{r}\frac{\partial}{\partial r} + \frac{1}{r^2}\frac{\partial^2}{\partial \phi^2} + \frac{\partial^2}{\partial z^2}\right)\psi(r, \phi, z) = E\psi(r, \phi, z) \tag{4.45}$$

$$-\frac{\hbar^2}{2m_p}\left(\frac{1}{r_0^2}\frac{\partial^2}{\partial \phi^2}\right)\psi(r_0, \phi, 0) = E\psi(r_0, \phi, 0). \tag{4.46}$$

Ignoring the constant r_0 for the moment, Equation 4.46 is the TISE in the single variable, ϕ, and it has the same general solution as the particle confined to the x-axis in the particle in a 1-dimensional box.

The solution is

$$\psi(r_0, \phi, 0) = Ae^{im\phi} + Be^{-im\phi}, \tag{4.47}$$

where[1] $m = \sqrt{\frac{2m_p r_0^2 E}{\hbar^2}} = \sqrt{\frac{2IE}{\hbar^2}}$ and $I = m_p r_0^2$ is the moment of inertia.

Once again, we can consider the first term to represent a particle moving clock-wise around the ring with a defined, positive angular momentum (see below), while the second term represents a particle moving counter-clock-wise around the ring with a defined, negative angular momentum. In either case, the pre-factor $(A$ or $B)$ can be determined by normalizing the wave functions to be $A = B = \sqrt{1/2\pi}$ so that

$$\psi(r_0, \phi, 0) = \sqrt{\frac{1}{2\pi}} \left(e^{im\phi} + e^{-im\phi} \right) = \sqrt{\frac{1}{2\pi}} \cos(m\phi). \tag{4.48}$$

From Postulate Ia we require continuity, which means that the wave functions must be periodic around the ring with the angle taking on values from $0 \leq \phi \leq 2\pi$, or $\psi(r_0, \phi, 0) = \psi(r_0, \phi + 2n\pi, 0)$ with n being an integer, which means that the quantum number m can have values of all positive and negative integers and in this case it can be zero as well.

We also note that the energy becomes

$$E = \frac{m^2 \hbar^2}{2m_p r_0^2} = \frac{m^2 \hbar^2}{2I}. \tag{4.49}$$

with $m = 0, \pm 1, \pm 2, \pm 3, \ldots$

The form of some of the wave functions on the ring (or circle) is shown in Figure 4.11 for some of the lowest set of quantum numbers.

As with the particle in the 1-dimensional box, the energies are quantized, depend on the square of the quantum number, depend inversely on the mass, and inversely on the square of the dimension of the box, here defined by the fixed radius of the ring.

The comparison with the 1-dimensional box with infinite walls is logical if one imagines the ring as arising from the line by bending it around so that the two edges meet. The analogy is complete if we consider the length of the box to be equal to the circumference of the ring, and if we recognize that only the even wave functions of the linear box will satisfy the continuity requirements on the circle (see problem assignment).

A key difference between the Hamiltonian in the linear coordinate and

---

[1] It is convention to use m as the quantum number for the angular momentum motion and it should not be confused with the mass of the particle, which we designate with a subscript as in m_p.

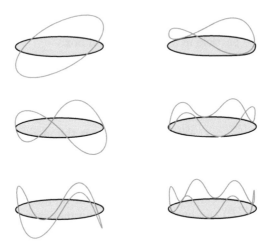

Figure 4.11: Examples of wave functions on a ring (or circle).

the circular coordinate is the incorporation of the radius of the ring into the pre-factor of the Hamiltonian. In analogy with the linear momentum operator, $\hat{p} = \frac{\hbar}{i}\frac{\partial}{\partial x}$, we can define an angular momentum operator for the motion around the z-axis, $\hat{L}_z = \frac{\hbar}{i}\frac{\partial}{\partial\phi}$, which allows us to compare the two Hamiltonians:

$$\hat{H}_{linear} = \frac{\hat{p}^2}{2m_p}\frac{\partial^2}{\partial x^2} \qquad (4.50)$$

$$\hat{H}_{angular} = \frac{\hat{L}_z^2}{2I}\frac{\partial^2}{\partial\phi^2}. \qquad (4.51)$$

In the linear case, the Hamiltonian operator contains the square of the linear momentum operator divided by the mass of the particle, while in the angular case, the Hamiltonian operator contains the square of the angular momentum operator divided by the moment of inertia of the particle. Just as the linear momentum is quantized, the angular momentum about the z-axis is quantized.

One key difference between the 1-dimensional box and the ring, is the probability distribution. In the ring, the probability is constant and independent of the position on the ring for all values of m:

$$\psi^*(r_0,\phi,0)\psi(r_0,\phi,0)d\phi = \sqrt{\frac{1}{2\pi}}e^{\mp imp}\sqrt{\frac{1}{2\pi}}e^{\pm imp} = \frac{1}{2\pi}d\phi. \qquad (4.52)$$

Thus, the particle is uniformly distributed around the ring for all the wave functions and for all energies. This is consistent with the uncertainty principle since the energies and momenta are precisely known, the position is not known at all as we saw earlier for the free particle. Note also that when $m = 0$, the energy is zero — there is a zero-point energy of zero. This is also possible since the position is not known at all.

Examination of the particle confined to a circle or a ring is instructive for a number of reasons.

1. It introduces the valuable concept that angular momentum is quantized. This finding leads us, as we will see later, to the concept of orbitals in the H-atom.

2. All the states where $m = -m$ are degenerate, so it does not matter which direction around the ring the particle is moving.

3. The system is a good application for electrons moving in circular conjugated systems, such as a benzene ring.

4. The solutions can be applied to particles confined to the surfaces of nanotubes, such as carbon nanotubes.

4.2.3.2 Confinement to a disk bounded by an infinite potential

The case where a particle is free to move within a circular disk of fixed radius r_0 and bounded by infinite potentials outside the disk is solved by the Hamiltonian Operator in cylindrical coordinates with the value of the z-coordinate being zero.

$$-\frac{\hbar^2}{2m_p}\left(\frac{\partial^2}{\partial r^2} + \frac{1}{r}\frac{\partial}{\partial r} + \frac{1}{r^2}\frac{\partial^2}{\partial \phi^2}\right)\psi(r,\phi,0) = E\psi(r,\phi,0) \qquad (4.53)$$

We use the fact that r and ϕ are independent coordinates and hence it is possible to separate the variables and obtain a wave function as a product of wave functions and the energy as sums of energies:

$$\psi(r,\phi,0) = \psi(r)\psi(\phi). \qquad (4.54)$$

The angular part of the wave functions was solved for any r-value for the ring to give

$$\psi_m(\phi) = \sqrt{\frac{1}{2\pi}}\left(e^{im\phi}\right) \qquad (4.55)$$

with $m = 0, \pm1, \pm2, \pm3, \ldots$

The radial part of the wave functions must now obey the radial Hamiltonian

$$-\frac{\hbar^2}{2m_p}\left(\frac{\partial^2}{\partial r^2} + \frac{1}{r}\frac{\partial}{\partial r} + \frac{m^2}{r^2}\right)\psi(r) = E_r\psi(r). \qquad (4.56)$$

This can be transformed into a well-known form of a differential equation, called a Bessel Equation of zero order:

$$\left(\frac{\partial^2}{\partial s^2} + \frac{1}{s}\frac{\partial}{\partial s} + \left(1 - \frac{m^2k^2}{s^2}\right)\right)\psi'(s) = 0. \qquad (4.57)$$

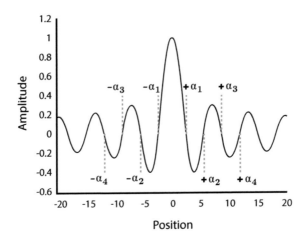

Figure 4.12: The zero-points in the $J_0(s)$ Bessel function.

where $s = kr$ and $k^2 = -\frac{2m_p E_r}{\hbar^2}$. This is in turn known to have solutions that are linear combinations of what are called Bessel Functions of the First, $J_n(s)$, and Second kind, $Y_n(s)$, both of order n such that

$$\psi'(s) = c_1 J_n(s) + c_2 Y_n(s). \tag{4.58}$$

The complexity of arriving at this solution is beyond the scope of this text, but for those interested, the transformation and the solutions can be found in Applied Mathematics or Engineering texts such as that by Kreyszig's *Advanced Engineering Mathematics*, where the problem is solved to understand the vibrations of vibrating planes, exemplified by objects such as drums. [See also Carl. W. David University of Connecticut Chemistry Educational Material http://digitalcommons.uconn.edu/cgi/viewcontent.cgi?article =1013&context=chem_educ.]

The Bessel Functions of the First and Second Kind are polynomial functions, and it turns out that $Y_n(s)$ is not bounded as s goes to zero (at the center of the disk) and is therefore not an acceptable part of the wave function. This leaves

$$\psi'(s) = c_1 J_n(s) = c_1 J_n(kr). \tag{4.59}$$

The polynomial functions describing the Bessel functions of the First Kind are decaying, oscillatory functions which are maximal when $s = 0$ and have zero values for a number of values of s (or kr) as exemplified in Figure 4.12 for $J_0(s)$ which corresponds to the case when the quantum number for the angular momentum, $m = 0$. The boundary condition for the problem requires that the wave function is zero at the edge of the disk (recall that Postulate Ia requires continuity and that the probability of finding the particle outside

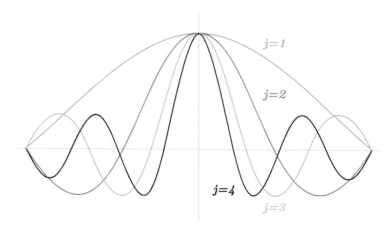

Figure 4.13: The four lowest energy radial wave functions for a disk

the disk where the potential is infinite is zero). This leaves us with a set of solutions, each corresponding to the section of the function up to a value where $J_n(kr_0) = 0$, designated $\pm \alpha_j$ as illustrated in Figure 4.12 for the quantum number $m = 0$ and with $j = 1, 2, 3, 4 \ldots$ being the corresponding quantum numbers for the radial function. The presence of the boundary condition leads to quantization of energies and wave functions.

Thus, in general

$$\psi_j(r) = c_1 J_n \left(\frac{\alpha_j}{r_0} r \right) \tag{4.60}$$

for $j = 1, 2, 3 \ldots$ and $n = 0, 1, 2, 3 \ldots$

In contrast to the solutions in the square box, where the wavelengths of the wave functions vary linearly with the quantum number, the wavelength for each wave function will be have a complex dependence on the quantum number. The wave functions are also known as the normal modes of the particle in the disk, corresponding to the normal modes of the vibrations of a drum. Figure 4.13 illustrates the first four normal modes along the radial direction for the case where $m = 0$. Note that the wave function has nodes along the radial direction where the wave function is zero and that the number of nodes is one less than the quantum number, as we also saw for the particle in the box in one dimension.

The total time-independent form of the wave function for the disk is now

$$\psi_{j,m}(r, \phi) = \psi_j(r)\psi_m(\phi) = c_1 J_n \left(\frac{\alpha_j}{r_0} r \right) \sqrt{\frac{1}{2\pi}} \left(e^{im\phi} \right) \tag{4.61}$$

for $j = 1, 2, 3 \ldots$ and $m = 0, \pm1, \pm2, \pm3, \ldots$

The energies for this system depend on the quantum numbers j and m (which in turn determines n). The simplest case is when $m = 0$ (and $n = 0$), for which $E_m = 0$ and which leads to

$$E_{r,j} = \frac{\hbar^2}{2m_p} \left(\frac{\alpha_j}{r_0} \right)^2. \qquad (4.62)$$

The wave functions and the energy levels can be visualized at the following site:

http://en.wikipedia.org/wiki/Vibrations_of_a_circular_membrane

---

The solution to the problem of a particle in a disk is interesting for several reasons:

1. It couples the radial part to an angular momentum component.

2. It illustrates that even though the details are more complex, the general principles observed in the rectilinear cases (lines, squares, and cubes) are retained, specifically:

    (a) The wave functions are oscillatory.

    (b) The energies are quantized.

    (c) The energy depends on the quantum number such that the energy increase as the quantum number increases (and the energy spacings increase as well).

    (d) The energy depends on the inverse of the mass of the particle, so as the mass increases, the energy level spacings become closer approaching the classical limit.

    (e) The energy depends on the inverse of the square of the radius of the disk, so we retain the general observation that the energy levels increase as the dimensions of the confinement decreases.

3. The disk is a reasonable approximation of a nanodot on a surface where the radius is greater than the thickness. This could for example be a material in contact between two cylindrical electrodes or a small patch of a metal in a sensor array.

4.2.3.3 Confinement to spherically symmetric systems: Spherical harmonics

The general form of the TISE in a spherically symmetric system uses the spherical coordinate system as expressed earlier so that we get

$$-\frac{\hbar^2}{2m_p}\left(\frac{1}{r^2}\frac{\partial}{\partial r}\left(r^2\frac{\partial}{\partial r}\right)+\frac{1}{r^2 sin\theta}\frac{\partial}{\partial\theta}\left(sin\theta\frac{\partial}{\partial\theta}\right)+\frac{1}{r^2 sin^2\theta}\frac{\partial^2}{\partial\phi^2}\right)\psi(r,\theta,\phi)$$
$$= (E - V(r))\,\psi(r,\theta,\phi). \tag{4.63}$$

Here we have kept the potential energy term to arrive at a general set of equations for the spherically symmetric systems.

Separation of variables leads to three interdependent differential equations expressed in terms of three components of the wave functions, $R(r)$; $\Theta(\theta)$;and $\Phi(\phi)$ that depend individually only on r,θ or ϕ:

$$\left(\frac{\partial^2}{\partial\phi^2}+m^2\right)\Phi(\phi) = 0 \tag{4.64}$$

$$\left(\frac{1}{sin\theta}\frac{\partial}{\partial\theta}\left(sin\theta\frac{\partial}{\partial\theta}\right)-\frac{m^2}{sin^2\theta}\frac{\partial^2}{\partial\phi^2}+l(l+1)\right)\Theta(\theta) = 0 \tag{4.65}$$

$$\left[-\frac{\hbar^2}{2m_p}\left(\frac{1}{r^2}\frac{\partial}{\partial r}\left(r^2\frac{\partial}{\partial r}\right)-\frac{l(l+1)}{r^2}\right)-(E-V(r))\right]R(r) = 0. \tag{4.66}$$

These are interdependent in the sense that they must be solved in the order listed. Nevertheless, the two angular equations can be solved generally without solving the radial differential equation, because neither depend on the variable potential. They become applicable to **all spherically symmetric systems** —these are the spherical harmonics.

Spherical Harmonics

As you will discover, the spherical harmonics are important functions that are used in many contexts. In order to stay consistent with the general use of these functions, we use the standard symbolism for these, $Y_l^m(\theta,\phi)$, and the functions that lead to them, $\Phi(\phi)$ and $\Theta(\theta)$.

There are numerous texts that show how these equations lead to the spherical harmonics and while relatively straightforward, it invokes raising and lowering operators related to the angular momentum operators, which is more detailed than needed for our purposes [for details see for example Herbert L Strauss *Quantum Mechanics: An Introduction.*]

Suffice it to say that they are defined as:

$$Y_l^m(\theta,\phi) = (-1)^m\left[\frac{2l+1}{4\pi}\frac{(l-m)!}{(l+m)!}\right]P_l^m(cos\theta)e^{im\phi}. \tag{4.67}$$

Here $P_l^m(cos\theta)$ is the Associated Legendre Polynomial which can be determined through known recurrence relationships (which means that if one is

Table 4.1: Spherical harmonics for $l = 0, 1, 2$

	$l = 0$	$l = 1$	$l = 2$
$m = +2$			$Y_2^2 = \sqrt{\frac{15}{32\pi}} sin^2\theta e^{2i\phi}$
$m = +1$		$Y_1^1 = -\sqrt{\frac{3}{8\pi}} sin\theta e^{i\phi}$	$Y_2^1 = -\sqrt{\frac{15}{8\pi}} sin\theta cos\theta e^{i\phi}$
$m = 0$	$Y_0^0 = \sqrt{\frac{1}{4\pi}}$	$Y_1^0 = \sqrt{\frac{3}{4\pi}} cos\theta$	$Y_2^0 = \sqrt{\frac{5}{16\pi}} (3cos^2\theta - 1)$
$m = -1$		$Y_1^{-1} = \sqrt{\frac{3}{8\pi}} sin^2\theta e^{-i\phi}$	$Y_2^{-1} = \sqrt{\frac{15}{8\pi}} sin\theta cos\theta e^{-i\phi}$
$m = -2$			$Y_2^{-2} = \sqrt{\frac{15}{32\pi}} sin^2\theta e^{-2i\phi}$

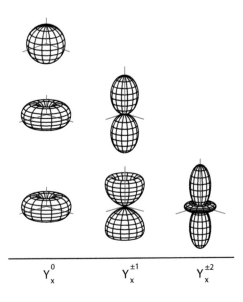

Figure 4.14: Illustrations of the spherical harmonics listed in Table 4.1.

known for a given m and l, those for $m \pm 1$ and $l \pm 1$ can be determined). It also is required that $|m| \leq l$ and $l = 0, 1, 2, 3, \ldots$. The values of m and l represent the quantum numbers for the two angles in the spherical coordinate system. The spherical harmonics for the values of $l = 0, 1,$ and 2 are shown in Table 4.1 and in Figure 4.14.

There are a number of significant aspects of these results:

1. These functions describe the angular part of the wave functions of all spherically symmetric systems.

2. The radial distribution of the potential will determine the details of the radial part of the wave functions as we will illustrate later.

3. The probability distributions are independent of the ϕ-angle.

4. The spherical harmonics are, as we shall discuss briefly later, responsible for the shape of the atomic orbitals in the H-atom model where $l = 0$ corresponds to s-orbitals, $l = 1$ corresponds to p-orbitals, $l = 2$ corresponds to d-orbitals. etc.

4.2.3.4 Confinement to a spherical shell by an infinite potential

Consider a particle of mass m_p confined to an infinitely thin, spherical shell of radius r_0 with a potential of zero in the shell and infinity elsewhere in space. This means that $V(r_0) = 0$ and hence

$$\left[\frac{\hbar^2}{2m_p} \left(\frac{l(l+1)}{r_0^2} \right) - E \right] R(r_0) = 0. \tag{4.68}$$

In other words, the radial wave function is a constant at $r = r_0$ and the energy is determined by

$$E_l = \frac{\hbar^2}{2m_p} \left(\frac{l(l+1)}{r_0^2} \right). \tag{4.69}$$

Note that the form of the energy is like that we have seen before for other problems:

1. The energy increases non-linearly with the increase in the quantum number as $l^2 + l$.

2. The energy is inversely proportional to the mass of the particle.

3. The energy is inversely proportional to the square of the radius of the shell to which it is confined.

This solution can, in principle, be applied to systems such as polymeric nanoparticles coated with a thin layer of a metal, or similar spherically symmetric layered nanoparticles.

The Rigid Rotator

It is interesting to note that the solution to the problem of a rigid rotator, which describes the possible motions of a diatomic molecule, is identical to the solution for the particle in a shell with a simple difference that the mass is replaced by the reduced mass of the diatomic molecule $\mu = \frac{m_1 m_2}{m_1 + m_2}$ and hence the rotational levels of a diatomic molecule is quantized (the angular momentum is quantized) and the difference between energy levels increases linearly, which gives rise to the characteristic multi-line rotational spectrum

$$\Delta E_l = E_{l+1} - E_l = \frac{\hbar^2}{\mu r_0^2} (l+1). \tag{4.70}$$

4.2.3.5 Confinement to a sphere bounded by an infinite potential

Consider finally the case where a particle of mass m_p is confined to a sphere of radius r_0 with an infinite potential outside. The angular part of the problem has been solved by the spherical harmonics and the radial part of the problem reduces to

$$\left[-\frac{\hbar^2}{2m_p} \left(\frac{1}{r^2} \frac{\partial}{\partial r} \left(r^2 \frac{\partial}{\partial r} \right) - \frac{l(l+1)}{r^2} \right) \right] R(r) = E_r R(r). \qquad (4.71)$$

We solved the equivalent problem for the radial part of the wave function with the disk, and the solutions will also be Bessel Functions of the First Kind, which, when subjected to the boundary conditions that the wave function is zero at the edge of the sphere, leads to quantization with energy levels similar to those seen in the case of the disk.

The details will not be discussed here, but it is evident that this system in principle can be applied to solid nanoparticles. In all cases, the energy levels will vary as the inverse square of the dimension of the particle.

4.3 PARTICLES CONSTRAINED BY FINITE POTENTIALS

The systems addressed in the previous chapter assumed that the particle of interest was confined to a region of space bounded by infinite potentials which ensured that the probability of the particle being outside the region of confinement is identically equal to zero. This in turn introduced boundary conditions that resulted in quantization of allowed energy levels corresponding to a set of wave functions that depend on quantum numbers, one for each dimension of confinement. These systems are particularly useful as illustrations of the general principles of confinement, namely the inverse dependence on the mass of the particle and the square of the dimension of the system irrespective of the detailed shape of the confinement. The systems are, however, not realistic since real potentials will be finite.

In this section, we will reconsider some of the systems we looked at in Section 4.2 but with finite, constant, potentials since this allows us to introduce some new concepts related to quantum phenomena, such as the ability for a particle to exist in regions of space where the potential exceeds the energy available to the particle; the ability of particles to tunnel from one region of confinement to another; the difference between bound and unbound states; and including the effect of the potential on the momentum of the particle.

In this chapter we will also consider a couple of systems in which the potential is a function of position to introduce the harmonic oscillator as a model for vibrations of molecular bond and the orbitals of the Hydrogen Atom.

4.3.1 Influences of barriers: Tunneling

In an earlier problem assignment, we demonstrated that if the particle is moving in a straight line (along the x-direction) in a constant, finite potential

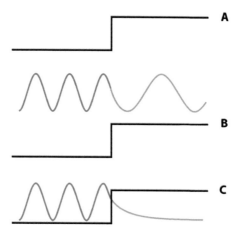

Figure 4.15: The effect of a particle approaching a barrier of constant height (A) when the energy is greater than the barrier (B) and when the energy is less than the barrier (C).

lower than the energy available to the particle, then the Hamiltonian is

$$-\frac{\hbar^2}{2m_p}\frac{\partial^2}{\partial x^2}\psi(x) + V_0\psi(x) = E\psi(x) \tag{4.72}$$

$$-\frac{\hbar^2}{2m_p}\frac{\partial^2}{\partial x^2}\psi(x) = (E - V_0)\psi(x). \tag{4.73}$$

Since V_0 is a constant, the solution of the wave function is the same as before

$$\psi(x) = Ae^{ikx} + Be^{-ikx} = \psi^+ + \psi^- = A'\cos(kx) + B'\sin(kx). \tag{4.74}$$

But the wavelength is increased since the total energy is lowered by the constant potential energy,

$$k^2 = \frac{2m(E - V_0)}{\hbar^2}. \tag{4.75}$$

Consider the case when the particle is moving with a positive momentum from a region of zero potential for $x < 0$ to a region of constant potential V_0 for $x > 0$ as illustrated in Figure 4.15A. We now need to further consider two scenarios: that where $E > V_0$ and that where $E < V_0$.

In the first scenario, the particle will continue to move forward with a positive momentum, but this is now reduced since the value of k changes from $k = \frac{\sqrt{2mE}}{\hbar}$ to $k = \frac{\sqrt{2m(E-V_0)}}{\hbar}$, which is still positive and real since $E > V_0$.

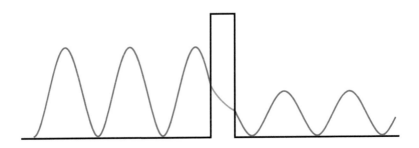

Figure 4.16: Illustrations of the effect of tunneling through a barrier of finite thickness.

Correspondingly, the wave function $\psi(x) = Ae^{ikx}$ is still represented as a sinusoidal wave but with a longer wavelength $\lambda = \frac{2\pi}{k}$ as illustrated in Figure 4.15B.

In the second scenario, the value of k becomes irrational since $k = \frac{\sqrt{2m(E-V_0)}}{\hbar} = \frac{\sqrt{-2m(V_0-E)}}{\hbar} = \frac{i\sqrt{2m(V_0-E)}}{\hbar}$. As a consequence, the wave function becomes $\psi(x) = Ae^{-kx}$, which for positive values of x is a real, exponentially decaying function within the barrier. The probability, $|\psi(x)|^2 = A^2 e^{-2kx}$, of finding the particle within the region of $x > 0$ is greater than zero, even though the energy available to the particle is less than the potential energy in that region, as illustrated in Figure 4.15C.

This finding is contrary to the classical picture. Classically, an object thrown at a wall will not penetrate into the wall unless the energy of the object is greater than the potential of the wall. Quantum mechanically, a particle approaching a wall has a finite probability of penetrating into the wall, and the probability of finding the particle inside the wall will decay exponentially with the distance into the wall. Note that as the potential increases, the probability decreases rapidly and in the limit where the potential is infinite, the probability is zero as expected from our earlier discussion in Section 4.2.

While the concept that a particle can exist in a region where the potential is greater than the energy of the particle is remarkable in its own right, it becomes even more interesting if we consider the barrier to be of finite thickness as illustrated in Figure 4.16. In this case, there are three regions of interest: $x < 0$, $0 < x < b$, and $x > b$ in which the potentials are zero, V_0, and zero respectively.

In the first region, the particle is described by the wave function $\psi_{x<0}(x) = Ae^{ikx}$ with a momentum given by $k = \frac{\sqrt{2mE}}{\hbar}$; in the second region, the particle is described by the decaying wave function $\psi_{0<x<b}(x) = Ae^{-kx}$ with a "momentum" given by $|k| = \frac{i\sqrt{2m(E-V_0)}}{\hbar}$; and in the third region the particle is again described by the wave function $\psi_{x>b}(x) = Ae^{ikx}$ with $k = \frac{\sqrt{2mE}}{\hbar}$. The particle, moving along the positive direction of the x-axis, penetrates the barrier and emerges on the other side of the barrier. Once there, it continues along the x-axis with the momentum, and hence the wavelength, of the particle unaltered. The particle has tunneled through the barrier without having enough energy to pass over it.

The probability that the particle emerges on the other side depends on the height, V_0 and width of the barrier, b, since $|\psi(b)|^2 = A^2 e^{-2kb}$ and $|k| = \frac{i\sqrt{2m(E-V_0)}}{\hbar}$.

The concept of tunneling is an important consequence of the postulates of quantum mechanics and is encountered in numerous chemical systems and has consequences for nanoelectronics as well. For example, if electrons are moving in a thin wire adjacent to a conductive surface, it is possible to have the electron escape from the wire by tunneling across the space between the wire and the conductive surface.

Tunneling from a thin wire tip to a surface, is also the foundation for the Scanning Tunneling Microscope (STM) which has been used to study the surface properties of materials at the atomic scale for a few decades. The invention of the STM led to a Nobel Prize in 1986.

4.3.2   One-dimensional boxes bounded by finite potentials

The understanding developed in the previous section of this chapter allows us to readily understand the system in which a particle is constrained to a one-dimensional box bounded by finite potentials as illustrated in Figure 4.17.

Inside the box with $V(x) = 0$ the solution to the wave function is the same as for the particle in the box with infinite walls.

$$-\frac{\hbar^2}{2m_p}\frac{\partial^2}{\partial x^2}\psi(x) = E\psi(x). \tag{4.76}$$

which has the same solution as for the free particle:

$$\psi(x) = Ae^{ik_0 x} + Be^{-ik_0 x} = \psi^+ + \psi^- = A^{'}cos(k_0 x) + B^{'}sin(k_0 x). \tag{4.77}$$

with $k_0^2 = \frac{2mE}{\hbar^2}$, but with different boundary conditions.

Outside the box with $V(x) = V_0$

$$\psi(x) = Ae^{ik_1 x} + Be^{-ik_1 x} = \psi^+ + \psi^- = A^{'}cos(k_1 x) + B^{'}sin(k_1 x). \tag{4.78}$$

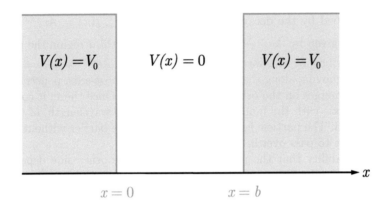

Figure 4.17: The one-dimensional box, or potential well, with finite walls.

with $k_1^2 = \frac{2m_p(E-V_0)}{\hbar^2}$.

The boundary conditions require that the wave functions must be continuous at the edges of the box.

We consider three cases separately: (i) $E < 0$; (ii) $E > V_0$; (iii) $0 < E < V_0$.

1. For $E < 0$, both k_1 and k_0 are imaginary, this means that ik_1 and ik_0 are both real numbers for all x, this in turn means that the wave function is not bounded as x tends to either plus or minus infinity. This in turn means that the wave functions do not exist and hence $E < 0$ is not possible!

2. For $E > V_0$, the solution is that of the free particle for all values of x, but the wavelength changes over the box!

    (a) Outside the box ($0 < x$ and $x > b$), $\lambda = \frac{h}{|p|} = \frac{h}{\sqrt{2m_p(E-V_0)}}$.

    (b) Inside the box ($0 < x < b$) $\lambda = \frac{h}{|p|} = \frac{h}{\sqrt{2m_p E}}$

    which corresponds to a smaller (shorter) wavelength or higher momentum of the particle.

    This case corresponds to unbound states of the particle. The effect of the presence of the box is that the momentum, or the wavelength, of the particle changes, but there is no quantization of the particle and it is free to move anywhere in space.[2]

---

[2]We can think about this as a pearl moving on a string where there is friction on both sides of a region defining the box but no friction inside this region. The pearl can move everywhere but will move more slowly outside than inside this region.

3. For $0 < E < V_0$, there are three wave functions to consider:

(a) $\psi(x) = Ae^{ikx} + Be^{-ikx} = Be^{k'x}$ for $x < 0$, which is bounded since $k' = \frac{\sqrt{2m(V_0-E)}}{\hbar}$ is real.

(b) $\psi(x) = Ae^{ikx} + Be^{-ikx} = Ae^{-k'x}$ for $x > b$, which is bounded since $k' = \frac{\sqrt{2m(V_0-E)}}{\hbar}$ is real.

(c) $\psi(x) = A'e^{ikx} + B'e^{-ikx}$ for $0 < x < b$,

which is now a sinusoidal solution in the box. The boundary conditions will leave us with four equations with four unknowns.

At $x = 0$ we can equate wave functions and their first derivatives from (a) and (c).

At $x = b$ we can equate wave functions and their first derivatives from (b) (and c).

The effect of these boundary conditions is to introduce quantization of energies as in the case of the one-dimensional box bounded by infinite potentials; the energies of these quantum levels are similar, but smaller, and depend once again on a quantum number squared, on the inverse of the mass and on the inverse of the square of the dimension of the box.

In addition, these boundary conditions show that the particle can exist with some probability outside the box, i.e., it can penetrate into the walls. Once again, as the potential increases the probability of penetration outside the box decreases.

This case corresponds to bound states of the particle. There are a limited number of quantum states within the box, depending on the dimension of the box and on the magnitude of the potential.

It is possible to show (Strauss) that the energy of the particle in a box of width $b = 2a$ bounded by finite potential V_0 is given by

$$E_n = \frac{\hbar^2}{2m_p a^2} \alpha_n^2 \qquad (4.79)$$

with $\alpha_n = k_n a$ being determined by solving

$$\alpha_n |sec(\alpha_n)| = \beta = \frac{\sqrt{2mV_0}}{\hbar} a \text{ for odd values of } n$$

and

$$\alpha_n |cosec(\alpha_n)| = \beta = \frac{\sqrt{2mV_0}}{\hbar} a \text{ for even values of } n.$$

For finite values of V_0, the value of $\alpha_n < n\pi/2$, but as $V_0 \to \infty, \alpha_n \to n\pi/2$, which is the solution for infinite potentials.

This simple case illustrates an interesting set of concepts. For example, the transition from bound to unbound states would be a model for ionization an electron in a bond of a molecule or of an electron from nanoparticle to which

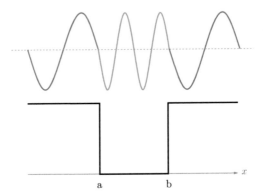

Figure 4.18: Illustration of the change in momentum (or wavelength) as a particle moves across a region of lower potential energy.

it is bound. One can also imagine two adjacent boxes with a barrier between them, through which the particle can tunnel from one box to the other.

Various examples of these cases are shown for $E > V_0$ in Figure 4.18 and for $0 < E < V_0$ in Figure 4.19. The latter case can also be explored further at http://www.falstad.com/qm1d/. The effect of changing the depth of the potential well is illustrated in Figure 4.20 and shows that the number of bound states depends on the barrier height. Correspondingly, the effect of changing the width of the potential well is illustrated in Figure 4.21 and it is clear that as the width decreases, the number of bound states decrease as well, since the spacing of the energy levels increases. The implication is that when the size of a nanoparticle decreases, the number of bound states decreases as well, all other things being equal.

4.3.3 Other systems bounded by finite potentials

In principle, all the systems considered in Section 4.2, namely two- and three-dimensional rectilinear boxes, rings, disks, shells, and spheres, can be solved with finite, rather than infinite potentials in a similar fashion. In all cases, there will be a distinction between unbound states, where the particle momentum will be influenced by the presence of the potential when it passes through that region of space, and bound states that will be quantized at energy levels that depend on the quantum numbers, the mass, and the dimensions of the system as we have seen before.

The solutions for two- and three-dimensional boxes will follow the pattern of the solution for the one-dimensional box, with the bound states being quantized at energy levels identical to those in the boxes with infinite walls; with a finite probability for the particle existing outside the box, with an exponentially decaying probability; and with the possibility of tunneling through po-

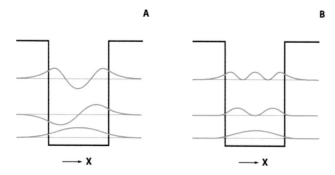

Figure 4.19: Illustration of the bound states for a particle trapped in a potential well. Note that the particle has a finite probability of penetrating into the region of high potential energy.

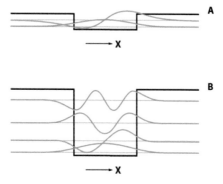

Figure 4.20: Illustration of the effect of changing the height of the potential well on the number of bound states. Note that the energy levels of the bound states are unaffected by the depth of the potential well.

tential energy barriers that exceed the available energy of the particle. Within the box, the bound state wave functions will still be oscillatory functions with nodes (where the particles cannot be found) that depend the on the quantum numbers of the state.

The solutions for rings and shells will also involve bound and unbound states, but for the bound states the angular portions of the problems will not change. Conceptually, however, there will be a finite probability that the particle in a bound state can exist at radial positions smaller and larger than

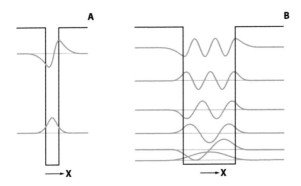

Figure 4.21: Illustration of the effect of changing the width of the potential well on the number of bound states. Note that the energy levels increase as the width decreases and hence the number of bound states decreases.

the fixed radius of the problem, r_0, with a probability decaying exponentially in either radial direction.

For disks and spheres, the angular portions of the problems will remain the same, and the radial portion will still be a described by the Bessel Functions as we saw earlier. However, the boundary conditions will change and require continuity of the wave functions and the derivatives of the wave functions at the edges of the disk and the sphere. These boundary conditions will alter the energy levels, but will also allow for a finite probability of the particle existing beyond the boundaries of the disk and the sphere.

The detailed solutions are beyond the purview of this text, but the conceptual findings arising from considering the particles bound by finite potentials are important: the particle is quantized, there are a finite number of bound states, there is a finite probability of the particle existing beyond the edges of the systems, and the particle can, in principle, tunnel from a bound state in one potential to a bound state in another potential. As we will see later, periodic potentials could represent a lattice structure where the valence electrons could tunnel from on lattice site to another.

The treatment of a particle in a potential would be incomplete without a brief discussion of three systems that are well known by chemists and form the model for understanding bonding and spectroscopy, namely:

1. The Hydrogen Atom, which is the simplest model for understanding atomic and molecular orbitals as well as electronic spectroscopy.

2. The harmonic oscillator, which is the simplest model for understanding vibrations of nuclei in diatomic molecules and the nature of vibrational spectroscopy.

3. The rigid rotator, which is the simplest model for understanding rotational motions of diatomic molecules and the nature of rotational spectroscopy.

4.3.3.1 The Hydrogen Atom

For the Hydrogen Atom the particle of interest is the electron with mass m_e and charge $-e$ in a potential determined by the Coulomb force between the electron and the nucleus of charge $+e$. Recalling that the Coulomb force between two charged particles is $F = \frac{-q_1 q_2}{r^2}$, then the potential in which the electron moves becomes $V(r) = \frac{Ze^2}{r}$ [for generality, we include the factor Z, which represents the nuclear charge and which is one for Hydrogen. Including Z allows us to generalize to "Hydrogen-like" systems, although these will always be approximations].

Without going through the detailed derivation (available in Strauss), one can show that the solution is

$$\psi(r, \theta, \phi) = R_{n,l}(r) Y_l^m(\theta, \phi). \tag{4.80}$$

Here the spherical harmonics control the shapes of the wave functions in the angular dimensions, so the orbitals are represented by the spherical harmonics as given in Table 4.1 and illustrated in Figure 4.10. The difference between these representations and the more familiar orbitals of $s; p_x, p_y, p_z; d_{xy}, d_{xz}, d_{yz}, d_{x^2-y^2}, d_{z^2}$ is that the latter have been rehybridized into the x, y, z-coordinate system.

The radial forms of the wave functions are given by

$$R_{n,l}(r) = \left(\frac{r}{a_0}\right)^l e^{\left(\frac{-Z}{na_0}r\right)} \sum_{k=0}^{n-l-1} d_k \left(\frac{Z}{na_0}r\right)^k. \tag{4.81}$$

Here a_0 is the Bohr radius, d_k are the coefficients of the power series that can be determined by a recurrence relationship in which

$$d_{k+1} = d_k 2 \left[k + l + 1 - n\right] / \left[(k+1)(k+2l+2)\right]$$

and d_0 is known from normalization. The quantum number, n, is the principal quantum number for which $n = 1, 2, 3, \ldots$ and $n > l(l = 0, 1, 2, \ldots)$.

These radial wave functions are decaying oscillatory functions as illustrated in Figure 4.22, where it can be seen that there are several nodes, including the nucleus, where the probability of finding the electron is zero.

The Energy is given by:

$$E_n = -\frac{\mu Z^2 e^4}{2\hbar^2 n^2} < 0. \tag{4.82}$$

There are several interesting points to note about the energy of these wave functions.

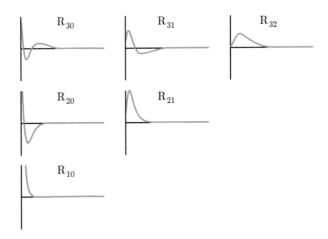

Figure 4.22: The radial wave functions for the H-atom.

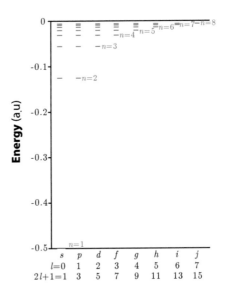

Figure 4.23: The energy levels of the H-atom (adapted from Strauss).

- First, the energy is negative! This arises because the Coulomb force, and hence the potential, is zero at infinite separations of the nucleus and the electron, and this is for simplicity the reference point. The attractive force increases as the separation decreases and hence the potential energy is negative. However, it has a lower bound when $n = 1$.

- Second, the energy is proportional to the reduced mass μ, which for all practical purposes is the mass of the electron ($\mu = \frac{m_p \cdot m_e}{m_p + m_e} \approx m_e$) since the mass of the proton is much larger than that of the electron).

- Third, in contrast to other cases examined so far, there is no explicit inverse dependence of the energy on the size of the confinement, here the orbital. However, the size of the orbital is indirectly incorporated into the quantum numbers, n and l, since the radial part of the wave functions depend on r^l. Hence as n and l increase, the size of the orbitals grow and the energies converge towards zero.

- Fourth, the energy depends inversely on the square of the principal quantum number n, which means that the energy separation between adjacent levels decreases as n increases, in contrast to the increasing separations observed in the earlier examples. These observations are evident from the energy level diagram in Figure 4.23.

- Fifth, the energy does not depend on the quantum numbers l and m, which means that these states are all degenerate.

4.3.3.2 The Harmonic Oscillator

If we consider two particles interacting through a linear force, such as a spring, then $F = -kx$ and the potential is a one-dimensional harmonic potential $V(x) = \frac{1}{2}kx^2$. There are two types of scenarios: one in which a particle is attached by a spring to a wall, which could be used as a simple model to calculate the vibration of an atom bound to a surface; or one in which two particles interact through a bond, such as two atoms in a diatomic molecule, which would be a simple model for the vibrations of atoms in a diatomic molecule. In either case, the form of the potential leads to the Hamiltonian

$$H = -\frac{\hbar^2}{2\mu}\frac{\partial^2}{\partial x^2} - \frac{1}{2}kx^2 \tag{4.83}$$

and the Time-Independent Schrödinger Equation to solve is

$$-\frac{\hbar^2}{2\mu}\frac{\partial^2}{\partial x^2}\psi(x) - \frac{1}{2}kx^2\psi(x) = E\psi(x). \tag{4.84}$$

where μ is the reduced mass. In the case of an atom, or particle, bound to a surface this is simply the mass of the atom, but in the case of the diatomic particle, it is the reduced mass that carries through.

There are a number of ways to solve this (available in Strauss), but in all cases there is a coordinate transformation from x to q such that $q^2 = (\mu\omega/\hbar)x^2$ in which ω is the angular frequency of oscillation of the classical harmonic wave related to the spring constant k as $k = \mu\omega^2$ (or $\omega = (k/\mu)^{1/2}$).

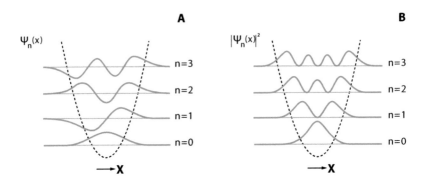

Figure 4.24: Wave functions, probability distributions, and energy levels for the simple harmonic oscillator.

The solutions to the Time-Independent Schrödinger Equation for this problem include the Hermite Polynomials, H_n:

$$\psi_n(q) = N_n H_n(q)e^{-q^2/2} \tag{4.85}$$

with

$$q = \left(\frac{m\omega}{\hbar}\right)^{1/2} x$$

$$N_n = \left(\frac{m\omega}{\hbar\pi}\right)^{1/4} \left(\frac{1}{(2^n n!)}\right)^{1/2}$$

$$H_n(q) = (-1)^n e^{q^2} \frac{\partial^n}{\partial q^n} e^{-q^2}$$

$$E_n = (n + \frac{1}{2})\hbar\omega. \tag{4.86}$$

The wave functions, the probability distributions and the energy levels for the lowest states are shown in Figure 4.24.

There are some key points to consider:

1. There is a single quantum number, $n = 0, 1, 2, 3, 4 \ldots$.

2. The wave function contains oscillatory components within the Hermite Polynomials, but these are damped by the exponential functions.

3. The Hermite Polynomials are not bounded, but the wave functions are because of the $e^{-q^2/2}$ term.

4. There are n nodes in the nth wave function.

5. The mass dependence is incorporated in the angular frequency, which depends inversely on the square of the reduced mass, hence the larger the mass, the smaller the energy separations and we will once again approach the classical limit for large masses.

6. The energy is linearly proportional to n, which means that all the energy levels are equally spaced [the separation between adjacent states are all equal to $\hbar\omega$].

7. Because of the evenly spaced energy levels, this is termed the harmonic oscillator and it serves as a first approximation for vibrational spectroscopy of diatomic molecules. In the ideal case, there will be only one vibrational excitation energy [$\hbar\omega$] and therefore only a single absorption line in the spectrum.

4.3.3.3 The Rigid Rotator

The rigid rotator is used as a simple model for the motion of two atoms bound by a fixed bond (i.e., no vibrations so that the separation of the atoms is fixed).

Consider first the general effect of the separation of variables. The angular parts are described by the spherical harmonics so we can now write the wave function as a product of wave functions

$$\psi(r, \theta, \phi) = R(r)\Theta(\theta)\Phi(\phi) = R(r)Y_l^m(\theta, \phi). \tag{4.87}$$

We can operate on this function with the Hamiltonian

$$\hat{H}R(r)Y_l^m(\theta, \phi) = \left[-\frac{\hbar^2}{2\mu r^2}\frac{\partial}{\partial r}\left(r^2\frac{\partial}{\partial r}\right) + \frac{1}{2\mu r^2}\hat{L}^2 + V(r)\right]R(r)Y_l^m(\theta, \phi)$$
$$= ER(r)Y_l^m(\theta, \phi) \tag{4.88}$$

which gives

$$\left[-\frac{\hbar^2}{2\mu r^2}\frac{\partial}{\partial r}\left(r^2\frac{\partial}{\partial r}\right) + \frac{1}{2\mu r^2}l(l+1)\hbar^2 + V(r)\right]R(r)Y_l^m(\theta, \phi)$$
$$= ER(r)Y_l^m(\theta, \phi). \tag{4.89}$$

Now the operator no longer has any angular components and it operates solely on $R(r)$.

For the special case of the rigid rotator (a diatomic molecule) of reduced mass, μ, the distance is fixed at the bond length r_0. Since $R(r) = R(r_0)$ is a constant (the bond length), and since $V(r) = 0$ [the potential is zero since they cannot move relative to each other, so there is no net force between them (Newton's third law)], then the total wave functions are simply the spherical harmonic wave functions scaled by a constant that depends on the distance separating the atoms. The TISE reduces to

$$\hat{H}R(r_0) = \frac{1}{2\mu r_0^2}l(l+1)\hbar^2 R(r_0) = ER(r_0) \tag{4.90}$$

which gives the energy

$$E_l = \frac{l(l+1)\hbar^2}{2\mu r_0^2}.$$

(4.91)

The energy of the rigid rotor depends only on the quantum number l and is independent of the quantum number m. There is degeneracy whenever $l > 0$.

As noted earlier, the separation in energies is linear in l, so the rotational spectrum for a diatomic molecule will have a series of absorption lines equally spaced in the energy.

CHAPTER 5

Applications to systems with many electrons

The preceding chapters provide a detailed mathematical description of the properties of a single particle constrained in space by potentials of various geometries. There are clear trends that emerge which allow us to think about the effect of quantum confinement. However, applications to more realistic systems require that we think further about systems with multiple electrons. Unfortunately, accurate descriptions of multi-electron systems are very complex because the electrons interact with each other. It is beyond the purpose of this text to deal properly with electron interactions, but we will consider some idealized systems that provide reasonable first approximations and illustrate the principles that we need to understand.

The objective of the current chapter is therefore to illustrate that the basic understanding of the quantum confinement principles can be extended to multi-electron systems in molecules, metals and semiconductors and their properties can be explored in a variety of manners. We will first consider simply filling the energy levels calculated in the previous chapter with more and more electrons and look at the consequences. This will lead to the concepts of bandgaps in metals and semiconductors. We will then examine the effects of having many electrons in a lattice structure and introduce some useful approximations that allow extension of the simpler forms of the Schrödinger Equation to complex systems. Finally, we will introduce the concept of photoluminescence as one of many tools to examine the properties of materials using electron transitions from bound to unbound states. At the end of the chapter, we will examine a few examples from recent literature that illustrate that theory and experiment are consistent, and while they differ in details, the properties of these systems can have interesting biological applications.

5.1 FILLING MULTIPLE LEVELS WITH NON-INTERACTING ELECTRONS

5.1.1 Conjugated electronic systems

Consider the simple one-dimensional particle in a box as a model for electrons in the π-orbitals of a linear, conjugated system of carbon atoms. In other words, let us model the π electrons in molecules such as ethene (ethylene), butadiene, hexatriene, octatetraene, etc., as electrons in a box with infinite potential outside the box and finite (for simplicity zero) potential inside the box, whose length is proportional to the number of conjugated double bonds. This was first done in 1952 by Bayliss who introduced it as a free electron approximation.

Ethene would have two electrons in the π orbitals, i.e., two electrons in the box. The size of the box is given to a the first approximation by the C–C bond length, which for ethene is $0.133\ nm$ (so $b = 133\ pm = 1.33 \times 10^{-10}\ m$). Postulate VI tells us that the electrons can occupy the same energy level if they have opposite symmetry, that is, they have different spin. We therefore expect the energy of the system to be given by Equation 4.25.

$$E_n = \frac{n^2 h^2}{8 m_e b^2} \tag{5.1}$$

The mass of the electron is $m_e = 9.10910^{-31}\ kg$; $h = 6.62610^{-34}\ J\ s$, so the lowest energy level, which contains the two electrons, would be for $n = 1$

$$E_1 = \frac{1^2 (6.62610^{-34})^2}{8(9.10910^{-31})(1.3310^{-10})^2} \frac{J^2\ s^2}{kg\ m^2} = 3.41 \times 10^{-18} J \tag{5.2}$$

This energy level is the highest energy level that contains electrons and we term it the Highest Occupied Molecular Orbital (or the HOMO Level).

The next lowest energy level is that for which $n = 2$ with four times (n^2) the energy: $E_2 = 1.36 \times 10^{-17} J$. This is the lowest energy level that contains no electrons and we term it the Lowest Unoccupied Molecular Orbital (or the LUMO Level). 1

If we apply energy corresponding to the difference between these two energy levels, we can promote an electron from the HOMO level to the LUMO level as illustrated in Figure 5.1. This energy can be provided by light at a frequency given by $\nu = \Delta E / h$ or a wavelength given by $\lambda = hc/\Delta E$. In this case, the wavelength is calculated to be $19.5\ nm$, which is a high energy excitation process. The experiment in which the electron in ethene is excited by light shows that the energy needed is actually a lot lower and corresponds to a wavelength of $180\ nm$ ($162.5\ nm$ in some sources in the gas phase). The particle-in-the-box model overestimates the energy significantly, but is within a factor of ten, which is not bad given the simplicity of the model.

By analogy, we can calculate the expected wavelengths for the longer conjugated systems. As we move from ethene to butadiene to hexatriene to octatetraene, we add two bonds – formally one single bond of length $148\ pm$ and

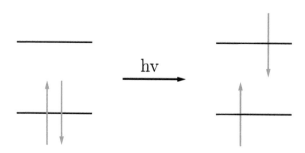

Figure 5.1: The principle of excitation of an electron from the HOMO to the LUMO in ethene.

one double bond of length 134 pm (or two bonds of average length 141 pm). In the process, we also increase the number of electrons in the conjugated system from 2 to 4 to 6 to 8, and hence the quantum number corresponding to the HOMO level increases from 1 to 2 to 3 to 4 (recall that there are two electrons in each energy level). The quantum numbers for the LUMO correspondingly increase from 2 to 3 to 4 to 5 and hence the excitation of the electron will occur at increasingly higher energy levels.

Table 5.1 summarizes the calculations of the wavelengths of excitation for these four molecules and compares them with the experiments. It is clear from these data that the particle-in-the-box model provides a reasonable estimate of the energy of the transition for the longer conjugated system but overestimates the energy of the transition for the shorter conjugated systems predicting much shorter wavelengths than observed. Nevertheless, the overall trend is the same, namely, the energy levels get closer together as the systems grow leading to lower wavelength excitation for longer conjugated systems. So, while the details of the calculations are imprecise, the qualitative predictions of the model is in accord with experimental observations.

While the one-dimensional particle-in-the-box is a simple model, it can be instructive to use it in combination with experimental data to calculate

Table 5.1: Comparison of Calculated and Experimental Wavelengths for Excitation of Electrons in Conjugated Systems

Compound	$b\ pm^{-1}$	n_{HOMO}	$E_n\ 10^{18}\ J^{-1}$	$\lambda_{calc}\ nm^{-1}$	$\lambda_{exp}\ nm^{-1}$
C_2H_4	133	1	3.41	19.5	180
C_4H_6	416	2	1.39	114	217
C_6H_8	698	3	1.11	230	258
C_8H_{10}	980	4	1.10	352	290

an "effective" size of the constraints placed on the electrons. To do so we use the observation (see problem assignment) that the size b is related to the excitation wavelength λ by

$$b = \sqrt{\frac{(2n+1)h\lambda}{8m_e c}} \tag{5.3}$$

thus for the four systems in Table 5.1, we would expect the effective lengths of the boxes to be 384 pm, 573 pm, 740 pm, and 890 pm respectively, somewhat longer than the simple addition of the bonds in the short systems and shorter in the long systems.

In the problem assignments we will explore some longer polyene systems, such as the carotenes as was also done by Bayliss.

5.1.2 Free electron gas model of metals

Metals readily conduct electricity, which can only be understood if the electrons associated with the individual atoms in the metal are bound less strongly than when in the free atoms. This implies that some of the valence electrons in the metal are free to move throughout the material.

Consider the progression of forming a metal from a metal such as Li or Na in which there is a single s-electron in the outer shell – the valence orbital

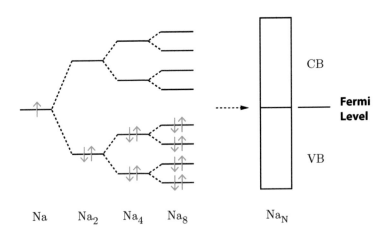

Figure 5.2: Illustration of electron distributions among the energy levels in progressively increasing sizes of clusters of a metal (Na), showing the formation of the valence band (VB) and the conduction band (CB) and the position of the Fermi Level

– as illustrated in Figure 5.2. That electron would be in the lowest energy level of a hydrogen-like orbital. Forming the Na-dimer, would create a linear combination of the atomic orbitals with one level being lowered relative to the atomic orbitals and one being raised by an equivalent energy. The two electrons would occupy the lower orbital, effectively forming a bond. Forming a tetrameric Na linear cluster would lead to further combinations of orbitals with two sets of orbitals lower than the original atomic orbitals and two sets of orbitals higher than the original atomic orbitals. The four electrons would occupy the two lower energy orbitals. An octameric cluster would have four sets of orbitals lower and four higher with the eight electrons occupying the four lower energy orbitals.

Note that as the clusters grow, the energies of the lowest orbital continue to decrease and hence the spread of available energies increases; the separation of the energies decreases or the "density of energy states" increases; the electrons all occupy lowered energy states leading to a more and more stable system; and the difference in energy between the highest occupied orbital and the lowest unoccupied orbital approaches zero. In the limit of very large clusters, the **density of states** is very high and the electrons occupy a nearly continuous range of energies.

In the ideal metal at low temperature, the electrons are all contained in states that cause the metal atoms to bond. They are in the **valence band of the metal**. The energy corresponding to the highest occupied orbital at zero degrees is called the Fermi Energy. All unoccupied orbitals correspond to energy levels that destabilize the systems and if an electron were to be excited above the Fermi Energy, it would no longer be bound and could move freely; it can conduct and hence the energies above the Fermi Energy correspond to the **conduction band of the metal**.

The free electron gas model, also called the Fermi gas model, for metals is based on a few simple assumptions:

- The solid metal lattice contains N positively charged nuclei and N negatively charged electrons which are free to move independently everywhere.

- The lattice structure is ignored.

- The Coulomb interactions are assumed negligible because of overall neutrality.

- Since the electrons are fermions, by Postulate VI, there are two electrons per energy level.

We can apply the results of the particle in the box with infinite walls and fill the energy levels with two electrons at a time. In the absence of thermal excitation the quantum number for the highest occupied energy level, corresponding to the Fermi Energy, would be $n_F = N/2$. In the case of the one-dimensional particle in the box, the Fermi Energy for the metal is

$$E_F = \frac{h^2}{8m_e} \left(\frac{N}{2b}\right)^2.$$ (5.4)

Note that the Fermi Energy is proportional to the square of the electron density (N/b)

The three-dimensional particle in the box is a better model and the energy is given by

$$E_k = \frac{h^2}{8m_e\pi^2}k^2 = \frac{h^2}{8m_e\pi^2}\left(k_x^2 + k_y^2 + k_z^2\right),$$ (5.5)

where k^2 is the square of the wave vector, $\mathbf{k} = \frac{2\pi}{b}(n_x, n_y, n_z)$, which in turn is related to the momentum, since $\mathbf{p} = \hbar\mathbf{k}$. Here we have assumed a cubic box of dimension b.

To get the Fermi Energy, we consider the system in the co-ordinate system of the wave vector. In that system, the energy levels will occupy a sphere of radius k_F which is the wave vector for the Fermi Energy level. This is most readily seen, as shown in Figure 5.3, where we see that

when $n_x = 1, n_y = 1, n_z = 1$ the energy level is at the coordinate $(1,1,1)$;
when $n_x = 2, n_y = 1, n_z = 1$ the energy level is at the coordinate $(2,1,1)$;
when $n_x = 1, n_y = 2, n_z = 1$ the energy level is at the coordinate $(1,2,1)$;
when $n_x = 1, n_y = 1, n_z = 2$ the energy level is at the coordinate $(1,1,2)$;

and likewise for (n_x, n_y, n_z) equal to $(3,1,1)$, $(1,3,1)$, $(1,1,3)$ and $(1,2,2)$, $(2,1,2)$, $(2,2,1)$ so that as the quantum numbers increase, the energy levels build in small blocks a solid body approximating a sphere. When all the electrons have been placed in energy levels, the last levels will be at the distance of k_F from the center.

The volume of this sphere is

$$V_{sphere} = \frac{4}{3}\pi k_F^3.$$ (5.6)

Likewise, the volume of an individual state in this coordinate system is

$$V_{state} = \left(\frac{2\pi}{b}\right)^3.$$ (5.7)

The number of electrons is N so the number of states is $N/2$, so the number of electrons can be expressed in terms of the ratio of the volume of the sphere and the volume of the states, i.e.,

$$N = 2\frac{V_{sphere}}{V_{state}} = \frac{b^3}{3\pi^2}k_F^3 = \frac{V_{sample}}{3\pi^2}k_F^3.$$ (5.8)

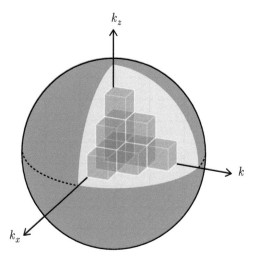

Figure 5.3: Illustration of the filling of the sphere in momentum space to get the volume of the Fermi sphere.

Correspondingly, the Fermi Wavevector can be calculated as

$$k_F = \left(3\pi^2 \frac{N}{V}\right)^{\frac{1}{3}},\tag{5.9}$$

which depends only on the electron density $\frac{N}{V}$, or on the density of states.
Finally, the Fermi Energy is

$$E_F = \frac{h^2}{m_e}\left(\frac{3}{4\pi}\frac{N}{V}\right)^{\frac{2}{3}}.\tag{5.10}$$

At higher temperatures, there will be a small number of electrons that move from the valence band to the conduction band and hence metals can conduct electrons quite easily. The distribution of electrons in energy levels around the Fermi Energy is given by Fermi–Dirac Statistics (see about distributions in later sections) and the fraction of electrons that a particular energy level contains is given by

$$f(E) = \frac{1}{exp[(E - \mu)/k_B T] + 1}.\tag{5.11}$$

Here μ is the chemical potential (see later sections about the thermodynamics). Note that as the temperature goes to zero, the occupation number for any energy level greater than μ approaches zero. Likewise, the occupation number for any energy level greater than the Fermi Energy approaches zero. It follows that the chemical potential approaches the Fermi Energy as the temperature goes to zero.

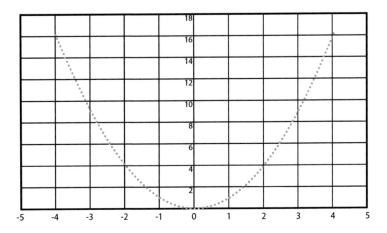

Figure 5.4: Illustration of the dependence of energy (y-axis) on momentum (expressed through k on the x-axis) for a nearly free electron in a crystal of fixed dimension.

5.1.3 Electrons in a lattice of atoms

While the Free Electron Gas model is a useful first approximation, it is not possible in general to ignore the size of the crystal or the effect of the atoms in the lattice.

5.1.3.1 Effect of crystal size

As we saw in Sections 4.2 and 4.3, the moment the electrons are confined to be within a specific region of space, the electron energies are quantized. That means that the parabolic dependence of energy on k is no longer a continuous dependence, but rather a series of discrete energies on the parabolic curve given by

$$E = \frac{\hbar^2 k^2}{2m}, \tag{5.12}$$

where k is given by

$$k = \frac{n\pi}{L} \tag{5.13}$$

and L is the dimension of the crystal and n is the quantum number with values of $1, 2, 3, \ldots$

This is illustrated in Figure 5.4 where the energy, E, is plotted as a function of the wave vector, k.

As the dimension of the crystal increases, the curve will be nearly continuous (the dots get closer), but if the dimension of the crystal decreases,

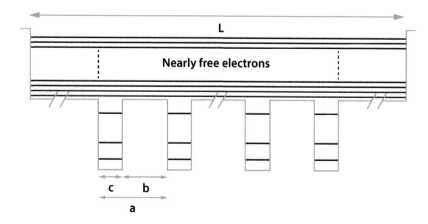

Figure 5.5: Potential of a 1D crystal of dimension L, with lattice spacing a, and potential wells at the atomic sites of width c and separated by the distance b.

the discrete energy states will become more prominent (the dots get farther apart).

5.1.3.2 Effect of the lattice structure

To illustrate the effect of the atoms in the lattice of the crystal consider the simple case of a one-dimensional lattice where the distance between atoms is given by a. Consider further that there is a rectangular potential well associated with each atom whose edges are separated by a distance b and the width of the well is c as illustrated in Figure 5.5

In this system we expect to see energy levels within the wells at the lattice sites to be discrete and given by the solution obtained in Section 4.3 for a square well with finite walls. Correspondingly, the energy levels of the nearly free electrons above these wells will be closely spaced and given to a first approximation by Equations 5.12 and 5.13.

However, the nearly free electrons will feel the effects of the periodic wells, and one must solve the Schrödinger equation for periodic boundary conditions, requiring that at the boundaries of the wells the wave functions and their first derivatives are continuous. This leads to a condition that

$$cos(ka) = \left(\frac{\alpha^2 - \beta^2}{\alpha\beta}\right) \left[sin\left(\beta\left(a-b\right)\right) sinh\left(\alpha\beta\right) + cos\left(\beta\left(a-b\right)\right) cosh\left(\alpha\beta\right)\right],$$

$$(5.14)$$

where

$$\alpha = \frac{\sqrt{2mE}}{\hbar}$$

$$\beta = \frac{\sqrt{2m(E-V)}}{\hbar} \qquad (5.15)$$

and $k = n\pi/a$).

This expression has only one variable, namely energy, E. Further, the expression requires that the energy is restricted to those values that in turn ensure the expression on the right hand side has values only between $+1$ and -1. The net effect is the emergence of bands of allowed energies and regions of forbidden energies, or gaps.

The effect of this set of restrictions on the energy is illustrated schematically in Figure 5.6, where the energies are now no longer confined to the parabolic curve (blue) but follow a different curve that is discontinuous[1] at the points in the curve where $k = n\pi/a$ as illustrated with the red curve. The group of allowed energies form a band of discrete energies while the forbidden energies form a gap between these bands as illustrated schematically with the blue boxes in the figure.

It is clear from this simple picture that the band structure depends on the lattice structure which will be different for every type of crystal lattice and will vary in the different directions in space depending on the symmetry of the unit cell in the crystal. In addition, the density of energy states within the bands depends both on the crystal structure and on the dimension of the crystal. In most cases, when the crystal is large, the allowed energies within the band will be nearly continuous, but we would expect that as the crystal gets smaller, there will be more discrete levels of the energy within each bands. As the crystal approaches dimensions on the nanoscale, one would qualitatively expect that this perturbation of the density of states could be significant.

It is also worth noting that since the band structure is sensitive to the crystal lattice, perturbations in the lattice, such as through thermal motion of the atoms, would affect the band structure and hence it will be possible to couple the momentum of the electrons to the motions of the lattice leading to an electron-phonon interaction, the phonon being a representation of the lattice movement.

In the diagrams in Figures 5.4 and 5.6 of energy plotted versus the wave vector, k, the positions along the k-axis corresponding to the discontinuities in energy, which is where $k = n\pi/a$, are called Bragg planes and

---

[1]Note that the discontinuity in energy is consistent with the solution in Section 4.3 where the energy (or wavelength) changed dramatically at the edges where the potential changes as illustrated in Figure 4.15.

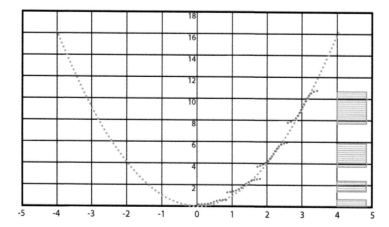

Figure 5.6: The appearance of the band structure as a consequence of interaction of the nearly free electrons with the periodic potential created by the lattice structure.

they represent the regions in the real lattice where electrons are diffracted. The regions between these Bragg planes are called Brillouin Zones. The first Brillouin Zone is the region where $k < \pi/a$ and the $n'th$ Brillouin Zone is the region where $(n-1)\pi/a < k < n\pi/a$. Because of symmetry and the repeat of the unit cells, it is possible to represent all the band structures in a plot of energy versus the wave vector within the first Brillouin Zone and that is the common pictorial presentation of band structures of materials. An excellent introduction to the concept of Brillouin Zones and band structures can be found in the lectures by Prathap Haridoss at IIT Madras available at http://nptel.iitm.ac.in.

It is important to note that the Fermi Energy is related to the chemical potential at zero temperature only. In order to understand the distribution of electrons in relation to the band structure, we need to define the Fermi Level. Recall we introduced the Fermi–Dirac Statistics through Equation 5.11. This represents the probability that a particular energy level is occupied by an electron. The Fermi Level is then defined as the chemical potential in this relationship and it follows that a state with energy equal to the Fermi Level will have a probability of 0.5 of being occupied.

If the Fermi Level falls within one of the bands, then the electrons are free to move and can conduct electricity and the material is termed a conductor. However, if the Fermi Level falls within one of the gaps, they are, in principle not able to conduct electricity and if the gap is large, the material is termed an insulator. If the gap is of intermediary levels, the thermal population may allow for limited conductivity, and if a potential is applied, then it can conduct and the material is termed a semiconductor.

5.1.3.3 An alternative approach

As we have seen, in real systems, the electrons interact with the atoms in each of the lattice positions since these will create a periodic potential, whose dimensions will depend on the orientation within the crystal lattice. We have seen that the effect of a non-zero potential is to reduce the effective momentum of the electron (or its wavelength).[2] While the momentum will vary periodically, on average over longer distances, there is a net reduction of the momentum, which could be treated as if the mass of the electron was reduced. This leads to the "effective mass" approximation, which allows us to use the simple form of the TISE for all materials with the mass of the electron substituted by the effective mass, m^*, for a given material. The effective mass would be obtained by fitting of experimental data to the expected energy values and can be either larger or smaller than the true mass.

In this way, we can retain the simplest form of the Schrödinger Equation along a particular direction, x, in the lattice:

$$-\frac{\hbar^2}{2m^*}\frac{\partial^2}{\partial x^2}\psi(x) = E\psi(x) \tag{5.16}$$

and the energies for the electrons are given by

$$E_k = \frac{\hbar^2}{2m^*}k^2. \tag{5.17}$$

The energy dependence on the wave vector (or momentum) given by Equation 5.17 is once again parabolic as shown in Figure 5.7. The effective mass approximation would then correspond to a broader parabolic function if the effective mass is greater, as illustrated in Figure 5.7. The width of the energy function will depend on the effective mass of the electron in a particular material and is therefore a reflection of the details of the lattice structure and of the strengths of the potentials in which the electrons move.

The energy curves in Figure 5.7 reflect the effect on the electrons that have energies greater than the Fermi Level i.e., those in the conduction band, rather than the electrons that are bound in the lattice to hold the lattice together. However, whenever an electron is promoted to the conduction band (thermally as discussed above or otherwise), there is a vacancy in one of the energy levels in the valence band. This vacancy is called a "hole" and is the counterpart to the electron in the conduction band. The holes in the valence band behave as an unbound positively charged particle within the valence band and hence we

---

[2]See Figure 4.15.

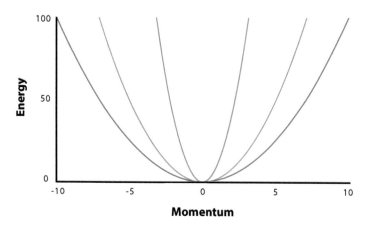

Figure 5.7: Illustration of the momentum dependence of energies in the conduction bands for different effective masses with the red curve corresponding to the true mass, the green to a smaller effective mass, and the blue to a larger effective mass.

can apply the same quantum mechanical treatment and use the same effective mass approximation. As a result, there will be an energy envelope equivalent to that in Figure 5.7, but it would be inverse since it is a "positive" particle.

For a metal, we would describe the valence band and conduction bands as illustrated in Figure 5.8A. This is an idealized figure with different effective masses for the holes and the electrons.

The details of the band gap structure will be anisotropic and will depend on the direction of the movement of the electron through the material.

5.1.4 Band structure of semiconductors and insulators

The simple picture forming clusters of metals as argued above and illustrated in Figure 5.2, is based on the linear combination of orbitals that lead to equal lowering and raising of energy levels because we start with essentially degenerate energy levels. This in turn gives rise to a band gap picture as in Figure 5.8A where the conduction and valence bands meet at zero momentum and at the Fermi Level because the energy of the highest occupied energy level is very close to the energy of the lowest unoccupied energy level. This picture is particularly relevant to the alkali metals (Li, Na, K, etc.) where it is a single s-electron that is involved.

Most other materials involve complex linear combinations of electronic valence orbitals, which in turn leads to more complex band structures. In general, the highest occupied energy level is at significantly lower energies

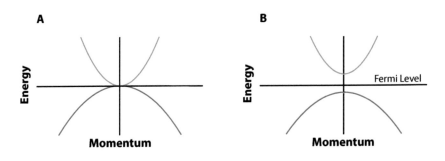

Figure 5.8: Illustration of the band gaps between valence and conduction bands in metals (A); semiconductors (B).

than that of the lowest unoccupied energy level. Nevertheless, the effective mass approximation can still be applied as an approximation and the energy dependence of both the valence band and the conduction band remains nearly parabolic.

The net effect is that there is an energy gap — called the band gap — between the maximum of the valence band and the minimum of the conduction band as illustrated in Figure 5.8B.

The Fermi Level will fall somewhere within this gap.

If the band gap is comparable to the thermal energy at room temperature, the material will act as a **conductor** since there will be some electrons in the conduction band.

If the band gap is much greater than the thermal energy but less than a few electron volts ($1\ eV = 1.60210^{-19}\ J$), then it is possible to populate the conduction band by applying a small voltage to the material and the material will conduct electrons in response to the applied electric field. These materials are called **semiconductors**.

If the band gap is much greater than several electron volts, the material will not conduct electrons even with a significant applied voltage. These materials are **insulators**.

In many, and perhaps most, cases, the electrons in the conduction band have a different dependence on the lattice structure than the holes in the valence band which affects the effective mass and hence the pitch of the parabolic function. Moreover, this difference can also lead to a shift in the minimum or maximum of the energy function. In that case the two band gaps do not align in the momentum space. This could be understood if the electrons in the con-

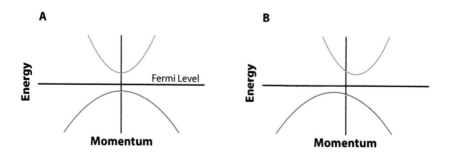

Figure 5.9: Illustration of the difference between direct (A) and indirect (B) band gap systems.

duction band lead to strain or relaxation in the lattice leading to a different structure or lattice dimension. This is illustrated schematically in Figure 5.9.

Those materials in which the maximum of the valence band and minimum of the conduction bands align in momentum space are called **direct band gap materials**, since it is possible to transfer electrons from one band to the other without changing their momentum. Correspondingly, those materials in which the valence and conduction bands do not align are called **indirect band gap materials** since the transfer of the electron from one band to the other requires a change in momentum, which in turn requires a coupling with the vibrational motions of the lattice; there needs to be a coupling with the phonons in the lattice. As we shall see, this leads to different optical properties of direct and indirect band gap materials.

5.1.5 Junctions between materials

The effective mass approximation assumes that the dimensions of the material is very large relative to dimensions of the lattice. At the edges of materials, the approximation fails, in part because the requirement for charge neutrality fails at the surface. The result is that the band structure will change — it bends. Likewise, when two different materials are juxtaposed to form an interface — a heterojunction — the band structure of the two materials will influence each other.

The concept of the heterojunction is illustrated in Figure 5.10 in several types of configurations. The effective mass approximation can be extended to the junction if the potential terms are included explicitly in the regions of

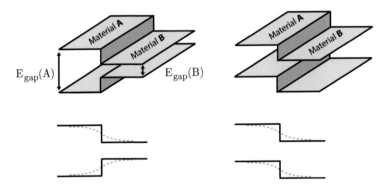

Figure 5.10: Schematic representation of different types of heterojunctions.

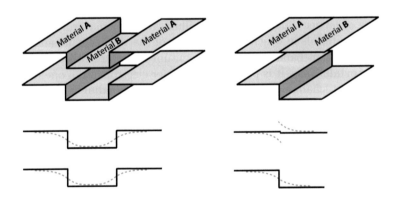

Figure 5.11: Creating a quantum well with a thin film of one material between two other materials.

relevance i.e.,

$$-\frac{\hbar^2}{2m^*}\frac{\partial^2}{\partial x^2}\psi(x) + V(x)\psi(x) = E\psi(x),\qquad(5.18)$$

which is the problem discussed in Section 4.4.1. for a barrier with the x-axis being chosen perpendicular to the junction.

Two heterojunctions in which a thin film of a material is included between two equal materials create what is called a quantum well as illustrated in Figure 5.11, which is also the problem discussed in Section 4.4.2, with the difference being that we invoke the effective mass approximation.

The extent of bending of the bands is controlled in part by the charge difference at the interfaces and will require that the wave functions are continuous at the edges.

The combination of the effective mass approximation and the application of the TISE with a potential change across the heterojunctions is also called the *envelope function approximation*. The terminology arises from the expectation that the physical properties are well represented by a slowly varying wave functions $\psi(x)$ without needing to account for the rapidly varying wave functions $\phi(x)$ which varies with x at the dimension of the lattice.

The use of the envelope function approximation (and inherently the effective mass approximation) allows for complex electronic structures to be amenable to computational exploration as has been well documented by Harrison.

5.1.6   Excitons

When an electron is promoted from a valence band to a conduction band in a semiconductor or an insulator, there is a Coulomb interaction between the negative electron and the positive hole that keeps them bound for a finite period of time forming an exciton. This interaction can be described in a manner similar to the interaction between the nucleus and the electron in the hydrogen atom (Section 4.3.3.1) and hence there will be a similar energy stabilization, although at a much smaller energy scale since the effective masses are smaller and since there are other charges in the region. The concept of the exciton is illustrated in Figure 5.12 both in the context of the valence and conduction bands and in the context of the lattice where the hole is associated with the atom of a specific lattice site.

At normal temperature there is a gradient of occupation of energy levels distributed around the Fermi Level given by the Fermi–Dirac distribution. Thus in a conductor there are always a significant number of electrons in the conduction band while in an insulator there are none. In the semiconductor the number of electrons in the conduction band is very low and they have to be promoted into the conduction band by some external factor. These situations are illustrated schematically in Figure 5.13

This figure also shows the effect of adding a dopant to the crystal. This is the process of adding a small amount of a different element that will fit within the basic crystal structure and that carries either more or fewer electrons. For example, if the crystal is that of silicon, then adding a small amount of phosphorous will introduce more electrons into the system, which causes the Fermi Level to be closer to the conduction band than to the valence band and there are more electrons in the conduction band. In contrast, if a small amount of boron is added to the silicon crystal, there is an electron deficiency

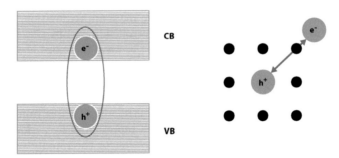

Figure 5.12: Illustration of the formation of an exciton relative to the band structure on the left and the crystal structure on the right.

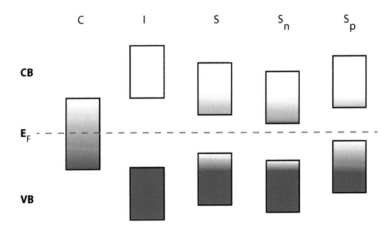

Figure 5.13: The density gradient around the Fermi Level (E_F) for conductors (C); insulators (I); and semiconductors (S) as well as for semiconductors doped to produce either an excess of electrons (S_n) or an excess of holes (S_p).

or a surplus of holes which means that the Fermi Level is closer to the valence band than the conduction band. Both of those situations are illustrated in Figure 5.13 by S_n and S_p respectively.

The shifting of the Fermi Level relative to the band structure is a useful tool in the nanoelectronics industry and it is particularly useful in junctions between n-doped materials and p-doped materials, which form the essential junctions in silicon based transistor circuits.

5.2   SELECTED EXAMPLES

The discussion presented here relies heavily on sources such as *Nanotechnology: A Crash Course* by R.J. Martin-Palma and A. Lakhtakia SPIE Press Bellingham, Washington USA 2010 (SPIE Digital Library eISBN 9780819480767; doi 10.1117/3.853406) and *Quantum Wells, Wires and Dots: Theoretical and Computational Physics of Semiconductor Nanostructures* by P. Harrison Wiley 2009 ISBN 9780470770986)

In general terms, Martin-Palma and Lakhtakia identify low-dimensional structures in terms of the number of degrees of freedom of motion available to a particle, as compared to the approach taken so far, in which we have referred to the dimensions in which the particle is constrained. Thus, in their terminology:

- Three-dimensional structures are bulk structures in which there are no constraint on the movement so the particle is free to move without any quantization effects.

- Two-dimensional structures, the quantum wells described above, are structures in which the particle is constrained in one direction, but free to move in two other directions. As we have seen, this leads to quantization in one direction only and can describe thin films of materials, typically embedded in other materials, with a single quantum number. This could also describe a quantum ring or tube.

- One-dimensional structures, also termed quantum wires, are structures in which the particle is constrained in two directions, but is free to move in the third direction. As we have seen, this leads to quantization in two directions and describe objects such as wires, long rods or pillars, or shells with two quantum numbers.

- Zero-dimensional structures, also termed quantum dots, are structures in which the particle is constrained in all three directions. As we have seen, this leads to quantization in three directions and describes objects such as spheres, cubes, or short rods with three quantum numbers.

As illustrative examples, consider the work described by Martin-Palma and Lakhtakia in which they consider the effect of confinement on electrons in materials such as Silicon (Si) and Gallium Arsenide (GaAs).

Their first example is the two-dimensional system of a 10-nm thin film of silicon (Si) illustrated in Figure 5.14 bounded by, first an infinite potential, and second by a finite potential of 200 meV. The four lowest energy levels for each system and the corresponding eigenfunctions are illustrated in Figures 5.15A and 5.15B.

> Their calculations are based on computational models discussed in detail by Harrison.

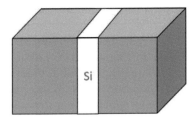

Figure 5.14: Illustration of a thin film of silicon imbedded between layers of other materials [adapted from Martin-Palma and Lakhtakia].

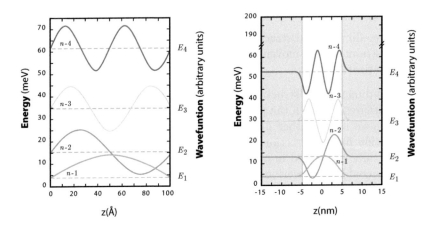

Figure 5.15: The first four eigenfunctions and energy levels for a 10 nm Si film bounded by an infinite potential (A) and finite potential of 200 mV (B) [adapted from Martin-Palma and Lakhtakia]

The calculation of these first four energy levels as a function of the width of the quantum well (or the thickness of the film) is shown in Figure 5.16. As expected, the energy levels decrease as the film thickness increases, as does the differences in the energy between the levels. However, in the ideal case where the solution to the Schrödinger equation is strictly correct, we would expect the change to be linear on a logarithmic plot like this with a slope of -2. The calculations (originally by Harrison) show that there are subtle variations from the ideal case. This does not change the general expectation as the size of the quantum confinement region increases, the energy level separations decrease and approach the classical behavior.

Their second example is a model of a quantum wire (Figure 5.17).

Their third example is a model of quantum dot (Figure 5.18).

These examples illustrate that while the details differ for different geome-

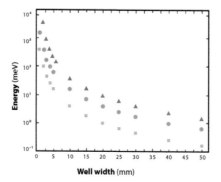

Figure 5.16: The first three eigenfunctions and energy levels for a 10-nm Si film bounded by finite potentials of 200 meV as a function of the width of the film [adapted from Martin-Palma and Lakhtakia].

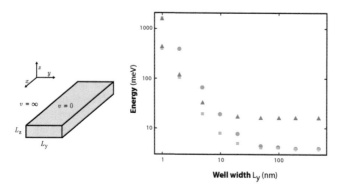

Figure 5.17: The first three eigenfunctions and energy levels for a 10-nm Si quantum wire bounded by finite potentials of 200meV as a function of the width of the wire [adapted from Martin-Palma and Lakhtakia].

tries, the general principle of quantum confinement leads to energy levels that depend on the dimensions of the confinement and as those dimensions increase, the differences in the energy levels decrease and approach the classical limit where there is no quantization of energy.

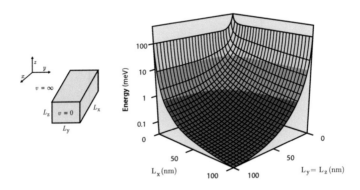

Figure 5.18: The energy level for a 10-nm Si quantum dot bounded by finite potentials of 200 meV as a function of the dimensions of the quantum dots [adapted from Martin-Palma and Lakhtakia].

5.3 PHOTOLUMINESCENCE IN SEMICONDUCTORS

In a later section, we will explore further the concept of absorption and emission of light in general. Here we consider the effect of emission of light from semiconductors as a tool to describe the properties of these materials.

In principle, an electron can be excited from the valence band to the conduction band by a number of sources of energy, such as thermal energy though heat, electrical energy through electric fields, or photons through absorption of light. In all cases, the result is formation of an exciton, which depending on the temperature, will either dissociate or relax. The minimal energy needed to create the exciton is the energy needed to overcome the band gap energy. Once formed, the exciton is stabilized by the Coulomb interaction between the positive hole and the negative electron. This system can be approximated as a Hydrogen-like system, with the hole acting as the nucleus and the energy is therefore lowered when they are close relative to their infinite separation. The energy E_C of the lowest state for that system is given by Equation 4.81 with $Z = 1$ and $n = 1$.

In bulk materials, the electrons and the holes formed upon excitation will have whatever excess energy is available during the excitation energy since both are free to move throughout the material.

In nanoscale structures, the electrons and the holes are formed in a confined environment and hence there is an additional minimal energy needed for each of them. This can be estimated by applying the relevant particle-in-a-box calculation to the electron and the hole, using the effective mass approximation. The semiconductor materials will absorb light at energies higher than the minimal energy needed, or at wavelengths shorter than that corresponding to the minimal energy. The minimal energy of this transition is the sum of the band gap energy (E_g); the lowest energy level for the electron in the conduction band (E_e as estimated from the particle-in-a-box calculation); the highest energy level for the hole in the valence band (E_h as estimated from the particle-in-a-box calculation); and the energy associated with the Coulomb interaction between the hole and the electron (E_C as estimated from a Hydrogen-like calculation with the zero-energy reference being the infinite separation of the electron and the hole). The difference in energy before and after the formation of the exciton is shown in Equation 5.19.

$$\Delta E = E_g + E_e + E_h + E_C. \tag{5.19}$$

The exciton is unstable and will eventually relax. The relaxation process will return the electron to the valence band and recombine with the hole to recover the original state of the system. During the relaxation process, energy must be released in the form of either thermal energy (such as enhanced lattice vibrations) or in the form of emission of electromagnetic radiation. The latter is termed **photoluminescence**. The efficiency of the photoluminescence will in turn depend on the relative competition between this process of relaxation and other competing processes.

In semiconductors that possess direct band gaps (Figure 5.19), the absorption and relaxation processes do not require any change in momentum of the electron or the lattice and the process of absorption and photoluminescence can be quite efficient. In contrast, in semiconductors with indirect band gaps (Figure 5.19), the electron must be excited to a higher energy state since the process of excitation of an electron is orders of magnitude faster than any possible movement of the nuclei in the lattice. The subsequent relaxation of the lattice from this lower energy state to the original state must then also involve coupling between the electrons and the phonons in the lattice. As a consequence, the efficiency of photoluminescence in indirect band gap materials is usually quite low.

A phonon is generated by vibrations of atoms in a lattice. There are, to first order, four vibrational modes that lead to phonons: (a) vibrations of adjacent atoms in phase (both move in the same direction) lead to acoustic phonons, which can either be along the crystal lattice direction (longitudial acoustic phonons) or perendicular to the crystal lattice direction (transverse acoustic phonons); and (b) vibrations of adjacent atoms out of phase (they move in opposite direction) lead to optic phonons, which can be along the crystal lattice direction (longitudial optical phonons) or perpendiculart to the crystal lattice direction (transverse optical phonons). The latter can couple with electromagnetic radiation since they cause relative movement of charges in space. Any actual vibrational motions can be decomposed as combinations of these four fundamental modes of movement.

Interestingly, many indirect band gap materials behave more like direct band gap materials when the dimension of the material decreases. For example, silicon is an indirect band gap material that does not exhibit photoluminescence in bulk materials, but nanoparticles of silicon can exhibit photoluminescence with high efficiency. Table 5.2 highlights a few band gap materials, their

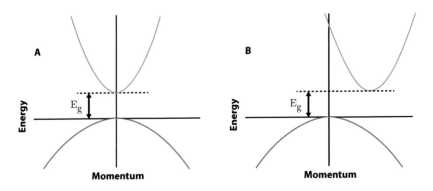

Figure 5.19: Direct band gap materials allow absorption and photolumines-cence processes without change in momentum (solid line in A) but indirect band gap materials require absorption into higher states (solid line in B) fol-lowed by relaxation to the lowest levels of the conduction band. Photolumi-nescence requires interactions with phonons and is therefore less likely.

Table 5.2: Bandgap Energies and Photoluminescence Wavelengths

Material	Band Gap Energy (eV)	Wavelength (nm)	Type of Band Gap
Si	1.11	1117	Indirect
Ge	0.66	1879	Indirect
GaAs	1.43	867	Direct
CdS	2.42	512	Direct
InP	1.27	976	Direct

band gap energy and the corresponding wavelength of the photon emitted during photoluminescence.

Yu and co-workers tested the implications of quantum confinement in In-dium Phosphide (InP) semiconductors wires and dots with the aim to compare these one- and zero-dimensional objects. They prepared nanowires of defined diameters and measured the absorption spectra of these structures. Some of their observed spectra are redrawn in Figure 5.20. It is evident that as the particle dimensions increase from an average of 3.49 to 8.84 nm, the wave-length of the photoluminescence increases from about 711 nm to about 842 nm (approaching 976 nm expected for the bulk material; see Table 5.2)

Based on application of the simple particle-in-the-box models they expect that the energy will vary inversely with the square of the dimension (d^2). Fig-ure 5.21 is adapted from their Figure 4 and confirms this trend. However, they

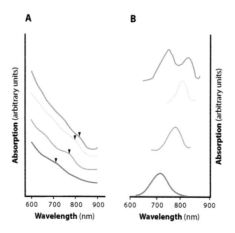

Figure 5.20: The absorption spectra of quantum wires of various diameters. Raw date at the left and background corrected data on the right. The four spectra corresponds to diameters ranging from about 3.5 nm to about 6.6 nm from the bottom to the top [adapted from Yu et al.].

found that the slopes are quantitatively incorrect and require a rescaling with the observation that the InP quantum wires and quantum dots are better described by a variation with $d^{-1.45}$ and $d^{-1.35}$ respectively. These are consistent with more advanced calculations and demonstrate that while the details may differ, the energy difference decrease as the dimensions increase.

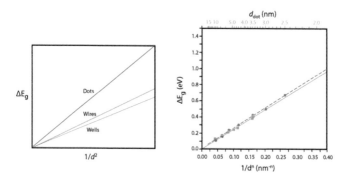

Figure 5.21: The band gap difference for InP quantum wires and dots of various diameters relative to the bulk [adapted from Yu et al.].

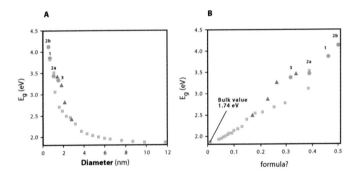

Figure 5.22: Variation in band gap energy for CdSe nanocrystals (assumed to be spheres) and CdSe molecular clusters (numbered) [adapted from Solviev et al.].

The fact that the simple particle-in-the-box model is inaccurate is not surprising and has been confirmed for other systems as well. For example, Ledoux and co-workers have examined the photoluminescence of silicon nanocrystals as a function of their size and find that the data is best fit to a first approximation to the relation

$$E_{PL} = E_0 + \frac{3.73}{d^{1.39}} + \frac{0.881}{d} - 0.245, \qquad (5.20)$$

where the last two terms allow for the fact that the surface differs from the bulk and hence the surface lattice parameters also depend on the dimension of the nanocrystals.

When does a molecule become a nanoparticle or vice versa? This question was examined for CdSe semiconductor materials by Soloviev et al. and represents one interesting example of addressing this transitional behavior. They prepared three CdSe molecular clusters whose structure is fully characterized by X-ray diffraction and whose sizes can be estimated accordingly. They used the photoluminescence spectra and, in particular, the photoluminescence excitation, to estimate the band gap energy for each of the compounds. They then compared these with the band gap energies estimated from the absorption spectra of a series of CdSe nanocrystals with diameters varying from 1.5 nm to 12 nm. Figure 5.22 illustrates the trend they observed as a function of diameter (A) and as a function of the inverse cubic root of the number of Cd-atoms in the system (B), which should be proportional to the inverse of the radius.

Two observations are clear — in the limit of very large sizes, the band gap approaches that of the bulk semiconductor band gap, and in the limit of very small sizes, there is good agreement between the band gap measurements of

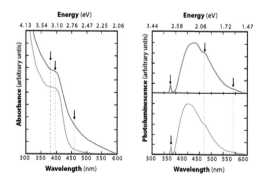

Figure 5.23: Absorption (left) and photoluminescence (right) from CdSe quantum dots in cultured HeLa cells [adapted from Silva et al.].

the molecular structures and the nanocrystals. In addition, the non-linear nature of the size dependence suggests that the band gap depends on the radius to a power between 1 and 2. In other words, the transition from nanoparticles to small molecular clusters appears to be rather smooth.

5.4   APPLICATION TO BIOLOGICAL SYSTEMS

Nanoparticles prepared from various semiconductor materials have become increasingly popular as imaging agents in biological systems. The applications are based on the very high photoluminescence yield that can be obtained and on the ability to tune the color of the luminescence through the size. The following examples are a few of the very large number of applications.

Silva and co-workers prepared CdSe nanoparticles — also termed quantum dots — with diameters of about 1.6 nm stabilized with a coating of thioglycerol (where the thiol reacts with the surface of the CdSe particle). They found that after six months, the absorption at around 460 nm grew as seen Figure 5.23 which provided evidence of the appearance of 1.6 nm particles. Over the same time period, the photoluminescence spectra changed to indicate the formation of Cd-Se structural features. These quantum dots were then incubated with HeLa cells in culture and they showed that the quantum dots could be visualized in the cells for as long as 36 hours after incubation, suggesting that they could be used for tracking cells as they move in culture or in tissues.

In optimized cases the intensity of the photoluminescence is sufficiently high that one can detect individual quantum dots[3] in cells which makes it possible to track them in real time as a function of time. Suzuki et al. provides

---

[3]Interestingly, in many cases the specific structure or chemical composition of the quantum dots used in biological applications is not known since it is proprietary information of the manufacturers

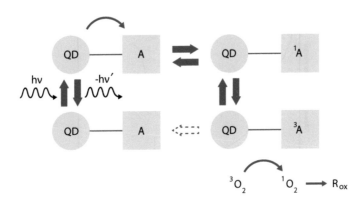

Figure 5.24: Schematic representation of the enhancement of photodynamic therapy by using quantum dots as the primary absorber [adapted from Biju et al.].

one example of this application in work where they followed the quantum dot attached to a small peptide which can promote translocation across cell membranes. As a result, they could follow the binding, translocation, and intracellular movement and determine that this could in turn activate the lateral transport and endocytosis processes.

It is possible to attach the quantum dots to proteins or other agents that will target them to bind to cancerous cells and hence accumulate in solid tumors. Because of their brightness, it now becomes possible to visualize the location of the tumors in whole animals by whole-animal imaging. Several examples are provided by Bentolila et al. in animals ranging from mouse models to zebra fish.

In later chapters, we will discuss in more detail the processes whereby electrons can be transferred from the excited state of a molecule to oxygen present in the surroundings to form what is known as reactive oxygen species — specifically, superoxide, hydrogen peroxide, and hydroxyl radicals. These species readily react with biologically active molecules within cells, causing cell toxicity. This can be used to target photoactive molecules to cancer cells, irradiate them, and cause specific, localized cell killing; this is known as photodynamic therapy and has been widely used with large porphyrin molecules that effectively absorb light and create reactive oxygen species effectively. Biju at al, have illustrated how this process can be enhanced by coupling these organic molecules to quantum dots. The quantum dots absorb the light, transfer the energy to the porphyrins which then create the reactive oxygen species as illustrated schematically in Figure 5.24.

III

Understanding surface
thermodynamics

Introduction to Part III

The properties of the atoms or molecules on a surface or at an interface differ from those of the atoms or molecules in the bulk primarily because the interatomic or the intermolecular interactions are different – for example, the number of nearest neighbors of like atoms or molecules is less at the surface.

The different properties of surface atoms and molecules have relatively little impact on the properties of the material as a whole if we measure these in large systems because the number of the distinct surface atoms or molecules is very small relative to the total number of atoms or molecules. For example, a cubic crystal of silicon measuring 1 cm by 1 cm by 1 cm contains 2.339 g or about 5×10^{22} atoms of Si. Depending on the crystal orientation, this same cube contains about 4.8×10^{15} atoms on the surface; in other words only about 1 in 10 million atoms are at the surface. However, a cubic crystal of silicon measuring 100 nm by 100 nm by 100 nm contains about 5×10^{7} atoms with about 4.8×10^{5} or about 1 percent of these on the surface. Thus for nanoscale structures with dimensions less than about 100 nm, the properties of the surface atoms or molecules will start to have a significant influence: the surface-to-volume ratio becomes large enough that the surface properties can become critical.

In order to understand the impact of the surface properties of nanoscale materials, we need to understand in more detail the thermodynamic properties of these materials. In this section the first chapter (Chapter 6) will provide a review of basic thermodynamics so that we understand the fundamental property of the chemical potential. The second chapter (Chapter 7) will examine in detail the effect of the size of the object on the chemical potential at the surface (the surface energy) and hence the impact on fundamental properties such as the equilibria and kinetics of reactions. The third chapter (Chapter 8) will examine how the surface properties affect the ease of removing electrons from materials and introduce the concept of work functions and will therefore link to the quantum mechanical properties discussed in Part II. The fourth chapter (Chapter 9) will be used to examine the interaction between the surfaces of different nanoscale objects and link the modern world of nanoscience and nanotechnology with the classical world of colloid science introduced well over a century ago.

CHAPTER 6

The basics of thermodynamics

6.1 SOME BASIC CONCEPTS

Thermodynamics refers to the field of chemistry which is concerned with understanding the flow of energy within a system or when a system changes from one state to another. One ultimate purpose of thermodynamics is to predict whether a change will occur spontaneously. The field of thermodynamics is based on three fundamental laws which lead to a set of parameters such as the free energy and the chemical potential, which are key to our understanding of spontaneous change.

In contrast to the field of quantum mechanics, the field of thermodynamics is, to a first approximation, not concerned with the detailed structures of the materials, and hence the outcomes become very generally applicable.

Our first task is to remind ourselves of some basic concepts of systems and how we describe these.

A **system** is simply that region of space or matter that we are interested in understanding. The **surroundings** of the system is everything else — in principle, everything else in the universe, but in practice everything else that matters, such as the laboratory in which we study the system. For example, the system could be a gas held in a glass container, with the surroundings being the glass container and everything around it. Similarly, the system could be a solution of molecules in a solvent where the system is the molecules in solution and the surroundings include the solvent, the container and everything around it.

We characterize a system by a number of functions (or variables) that tell us something unique about the system. Examples would be the temperature of the system, the pressure of the system, the volume of the system, the mass (or the number of molecules) in the system, and the energy of the system. If the state of the system changes, we expect one or more of these functions to change as well, for example, if the number of molecules in the system changes, the volume might change; or if a chemical reaction occurs, the temperature might change. In the context of thermodynamics we are most interested in understanding the change in energy, which is in turn often related to the changes in temperature, pressure, volume, and mass.

If changing a system to a different state and returning it to its original state causes the function (or variable) to return to its initial value, then that function (or variable) is termed a **state function** (or **state variable**). In other words, the function (or variable) describes the state of the system independently of how that state of the system is arrived at. This concept of path independence of state functions is important and will be used repeatedly.

For any given system there is a minimum number of independent state functions that will completely characterize the system and these are interrelated through an **equation of state**. A familiar example is the ideal gas law which interrelates the pressure, P, of a system, its volume, V, its temperature, T, and the number of gas molecules present, N:

$$PV = kNT. \tag{6.1}$$

Even though there are four variables associated with the system, the equation of state defines the fourth if three of them are known.

In general, we concern ourselves with three types of systems (see Figure 6.1):

1. **isolated systems**, which are systems that **cannot** exchange energy or mass with its surroundings.

2. **closed systems**, which are systems that **can** exchange energy but **not** mass with its surroundings.

3. **open systems**, which are systems that **can** exchange energy or mass or both with its surroundings.

Isolated systems are rare in practice but useful from a theoretical perspective and the Laws of Thermodynamics are described in terms of isolated systems. A gas or a liquid in a perfect, black thermos might be an example. Closed systems are quite common and allow us to examine exclusively the effect of energy flow between the system and the surroundings. Any container that prevents matter from entering or leaving, such as a flask with a stopper, represents a closed system. Open systems are the most complex since we need to keep track of both energy and mass flow, but also most interesting since they correspond to many real systems of interest. We will return to the differences between isolated, closed, and open systems in a later section where we discuss the effect of small numbers in nanoscale systems.

The path whereby a change of state occurs can be restricted such that one of the key state functions remains constant throughout the process. This is useful conceptually as well as experimentally, since it allows for controlled changes in state during which the number of variables is reduced. Some common processes are:

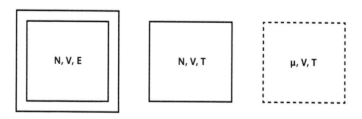

Figure 6.1: Schematic illustrations of isolated, closed, and open systems.

1. **isothermal** processes in which the temperature of the system is held constant,

2. **isobaric** processes in which the pressure of the system is held constant,

3. **isochoric** processes in which the volume of the system is held constant,

4. **isentropic** processes in which the entropy (see later) of the system is held constant, and

5. **adiabatic** processes in which there is no heat flow to or from the system.

Isothermal processes require that our system is in a large bath held at constant temperature that can transfer heat to or from the system in order to keep the temperature constant. Isobaric processes are typically those that occur in a container that can change volume to allow the internal pressure to always equilibrate with the external pressure. Open systems usually allow for isobaric processes. Isochoric processes occur in containers that cannot expand or contract so that there is no relationship between the external and internal pressure. Isentropic processes will be described later when we have defined the concept of entropy, but are generally more difficult to design. Adiabatic processes require that there is no heat flow to or from the system, but energy can still be exchanged in terms of work, so pressure, volume, and temperature can all change. This could happen in a thermos with a piston, or will happen in very fast processes where there is no time to exchange heat, such as the expansion of a gas that escapes from a balloon.

The process whereby a change in state occurs can be **reversible** or **irreversible**. Reversibility refers to the concept that at any point during the process of change, the process can be reversed along the exact path that it has taken so far. A reversible process usually involves a series of incremental steps while an irreversible process is usually an abrupt change. Consider the

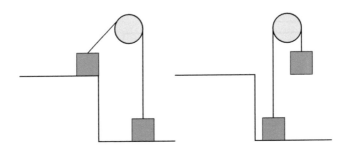

Figure 6.2: Illustration of energy as latent work.

analogy of descending from the side of a cliff to the bottom. A set of stairs offer a reversible path, while a jump represents an irreversible path.

With these concepts at hand we can now tackle the three laws of thermodynamics and understand some of the consequences of the laws. The approach taken here is analogous to that presented by PW Atkins in many of his physical chemistry texts.

6.2 THE LAWS OF THERMODYNAMICS

Energy is latent work. As illustrated in Figure 6.2, a rock at the top of a cliff has potential to do work, it has potential energy, since it can be connected through a pulley to a load such that when it is lowered from the top of the cliff it can perform work, here by moving a mass, m, by a distance x. The system can change energy by doing work on its surroundings or by the surroundings doing work on it.

Energy is also latent heat. For example, if we heat a fixed amount of a gas confined to a fixed volume, the temperature increases, and by the equation of state, the pressure must increase correspondingly. The increased pressure means that the gas has an increased capacity to do work, so the energy must have increased as well (Figure 6.3). The system can change energy by transferring heat to the surroundings or by the surroundings transferring heat to the system.

A system can change energy through work, w, or heat, q, and hence change from an initial energy, U_i, to a final energy, U_f, through any combination of work and heat[1] and we therefore expect

---

[1]In general a system can also change energy by other processes such as radiation or absorption of light or electric or magnetic energy, but it is always possible to convert these types of energy to a form of work or heat, so the following arguments remain general.

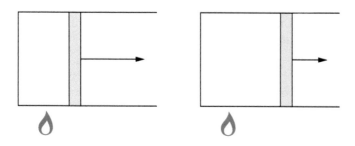

Figure 6.3: Illustration of energy as latent heat.

$$U_f - U_i = \Delta U = w + q. \qquad (6.2)$$

In this text we will designate the internal energy in the thermodynamic context with the symbol U. In other texts the internal energy is given the symbol E, which we in Part II used to describe the energy associated with quantum mechanical systems. In Part IV we will revert to using E for energy when we consider small molecular systems. The distinctions are fairly arbitrary. We also follow the convention that state functions are designated with uppercase letters while non-state functions such as work and heat are designated with lowercase letters.

There are, in principle, a number of ways in which the system can change from the initial state to the final state, ranging from the extremes of doing no work to transferring no heat, or any combination in between. Neither work nor heat are functions of the state of the system and both depend on the process of moving from the initial to the final state. However, their sum is a state function, which is independent of the path.

Consider a system changing from an initial state to a final state by two different paths, an adiabatic path in which there is no heat transfer ($q = 0$), and a non-adiabatic path in which there is heat transfer. Comparing the two, the change in energy is the same, so

$$\Delta U = w_{adiabatic} = w_{non-adiabatic} + q$$
$$q = w_{adiabatic} - w_{non-adiabatic}. \qquad (6.3)$$

which provides a mechanical interpretation of heat (see Atkins for more elaborate discussions of this principle).

6.2.1 The First Law

There are numerous expressions of the First Law of Thermodynamics. The most concise relates to the energy of an isolated system. An isolated system cannot, by definition, exchange energy with the surroundings, so it can do no work and transfer no heat, so

the energy of an isolated system is constant or

$$\Delta U^{isolated\ system} = 0. \tag{6.4}$$

An analogous statement of the First Law of Thermodynamics for a closed system, which can exchange energy through work and heat, is that

infinitesimal changes in energy of a closed system, dU, is the sum of the infinitesimal *work done on or by* the systems, dw, and the infinitesimal *heat transferred to or from* the system, dq, so that

$$dU^{closed\ system} = dw + dq. \tag{6.5}$$

Note that by convention the work done **on** the system is positive, raising the internal energy, and the work done **by** the system is negative, lowering the internal energy. Correspondingly, the heat transferred **to** the system is positive, raising the internal energy, and the heat transferred **from** the system is negative, lowering the internal energy.

6.2.1.1 Some consequences of the First Law

Work The work performed on or by a system can take many forms (mechanical, electrical, magnetic) but in all cases can be reduced to the equivalent of moving an object a certain distance against a fixed force, i.e., some form of mechanical work. Figure 6.4 illustrates this concept by raising a weight by a distance $\Delta z = z_f - z_i$ against the gravitational force $-F_g$ (here negative since it acts in the downward direction while the movement is in the upward direction).

The work done **by** the system in this case is

$$w = \int_{z_i}^{z_f} -F_g dz. \tag{6.6}$$

In general, if the force depends on position, the work done by the system is

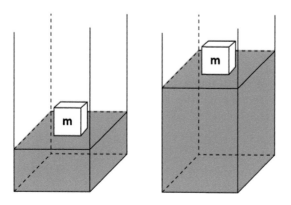

Figure 6.4: Illustration of work done by a system.

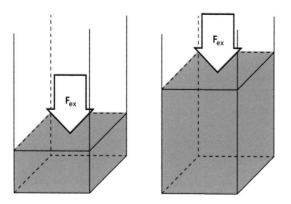

Figure 6.5: Illustration of expansion against an constant force

$$w = \int_{z_i}^{z_f} F(z)dz. \qquad (6.7)$$

In the case of a gas expanding against an external force, F_{ex} acting on an area A, as shown in Figure 6.5, the work done on the system is

$$w = \int_{z_i}^{z_f} -F_{ex}dz = \int_{z_i}^{z_f} (-F_{ex}/A) \times (Adz) = -\int_{z_i}^{z_f} P_{ex}dV. \qquad (6.8)$$

Here, P_{ex} is the external pressure (a positive quantity). This type of work is termed "pressure-volume work" or "PV-work".

When the volume decreases, the work is positive, so work is done on the system and contributes to an increase in internal energy. Correspondingly, when the volume increases, the work is negative, so the system does work on the surroundings and causes a decrease in the internal energy.

Heat   Heat can be transferred to or from a system resulting in a change in temperature of the systems. For small changes, the relationship between the heat transferred, dq and the change in temperature, dT is linear and the proportionality constant is called the **heat capacity**, C, so

$$dq = CdT. \tag{6.9}$$

For **isochoric** processes in a closed system, the volume is constant so the pressure-volume work is zero and by the First Law, and without any other type of work

$$dU = dq_V = C_V dT \tag{6.10}$$

or, in general, the **heat capacity at constant volume** in the absence of non-PV-work is

$$C_V = \left(\frac{dU}{dT}\right)_V. \tag{6.11}$$

For an **isobaric** process in a closed system, the internal pressure is constant and hence the PV-work is finite. By the First Law, the heat transferred at constant pressure is

$$dq_P = dU + PdV = C_P dT. \tag{6.12}$$

where we have introduced the **heat capacity at constant pressure**, C_P as the proportionality constant between heat transferred and temperature change.

Equation 6.12 suggests that there may be another convenient expression for energy that is the sum of the internal energy and the energy associated with the product of P and V.[2] This new expression for energy is called **enthalpy**, H, and is defined as

$$H = U + PV. \tag{6.13}$$

It is easy to show using the total differential of PV that

$$dH = dU + PdV + VdP \tag{6.14}$$

and that at constant pressure $(dP = 0)$

$$dH = dU + PdV = dq_P = C_P dT, \tag{6.15}$$

---

[2]Note that the units of PV are those of energy.

where we see the relationship between enthalpy change and the heat capacity at constant pressure.

Finally, the heat capacity at constant pressure in the absence of non-PV-work becomes

$$C_P = \left(\frac{dH}{dT}\right)_P.$$ (6.16)

Both the internal energy U and the enthalpy H represent measures of the energy of the system of interest. We use both since the internal energy change is a direct measure of the heat transferred during an isochoric (constant volume) process while the enthalpy change is a direct measure of the heat transferred during an isobaric (constant pressure) process. We are, therefore, beginning to separate our thermodynamic discussions into two domains — those related to constant volume processes and those related to constant pressure processes. This distinction will continue later.

6.2.2 The Second Law

The Second Law of Thermodynamics is based on the concept of **entropy**, which is perhaps the most difficult concept to understand fully in the field of physical chemistry. Yet, it is critical, so we will spend some effort developing it.

6.2.2.1 *The concept of spontaneous change and entropy*

There are some processes that occur spontaneously and some that do not. A rock will spontaneously roll down a hill, but will not spontaneously roll up the hill, as is only too well exemplified by the legend of Sisyphus. Likewise, a gas will spontaneously escape from a balloon, but will not spontaneously fill the balloon. In the first case it is evident that there is a decrease in the potential energy of the rock, and in the second case it is evident that the potential to do work has decreased, so the energy of the system has decreased. We might believe that the decrease in energy is the driving force for spontaneous change.

However, consider an isolated system of gas molecules which initially have all the gas molecules located in one half of the box (Figure 6.6). We expect that these will spontaneously fill the whole box even though there is no change in internal energy as a consequence of the change (by the First Law ΔU is zero, there is no work done and there is no change in temperature). Likewise we will never expect to see all the gas molecules spontaneously gather in one half of the box if they are initially everywhere. We expect that we would have to do work on the system to compress the gas molecules to one part of the box. Thus we can have spontaneous change without a change in internal energy, so that cannot be a sufficient criterion. However, we can reverse the direction of spontaneous change by doing work.

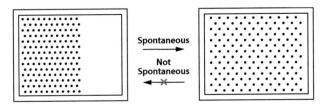

Figure 6.6: Spontaneous expansion of a gas to fill a volume.

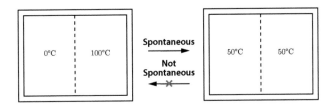

Figure 6.7: Spontaneous transfer of heat from hot to cold regions.

Similarly, consider two equal blocks of metal — one cold and one hot — in contact with each other but isolated from the rest of the surroundings (Figure 6.7). We expect that the cold block will spontaneously get warmer and the hot will get colder until they are at the same temperature. Conversely, we would never expect that the cold will spontaneously get colder and the hot will spontaneously get hotter, or, after the two blocks are at the same temperature that one will spontaneously warm up while the other spontaneously cools down. The overall internal energy of the two blocks has not changed since they are isolated, so there must be another criterion driving the spontaneous change.

The key difference between the two states of the gas in the first example is not the total internal energy of the system, but how that energy is distributed. When the gas occupies the whole box, it distributes the energy more widely and the system is less ordered. Likewise, the difference between the two states of the metal blocks in the second example is how the internal energy is dis-

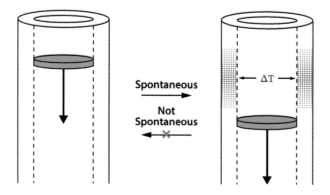

Figure 6.8: Spontaneous processes can transfer heat to the surroundings

tributed between them. A criterion for spontaneous change then appears to be related to creating a less ordered distribution of the total energy.

The spontaneous processes described so far are all examples of irreversible processes and since the initial and final states are different, there must be a state function that describes the change — **this state function we will call entropy**. By our logic, the entropy of the irreversible process has changed.

If the change is reversible, it is at all times possible to revert to the original state and there is no direction of spontaneous change. Hence, in isolated systems, *there will be no change in entropy during reversible processes*.

Consider a system which can change energy and in the process transfer heat to the surroundings, for example a block of a heavy material falling in a tube with friction against the walls (Figure 6.8). The change in height lowers the internal energy of the system without doing any work since there is no counteracting force. However, the friction will heat the surrounding walls. The spontaneous process is irreversible since we do not ever expect the block to rise even if we transfer heat from the surrounding walls to the block. This experiment suggests that the change in the entropy of the system may be related to the amount of heat transferred to the surroundings. Moreover, we might expect the change in the surroundings to depend on the initial temperature of the wall; for example, the temperature change in the wall may be greater at lower temperature. This suggests that the change in entropy will be related to the temperature of the surroundings. As a consequence we **define the change in entropy of the surroundings as**

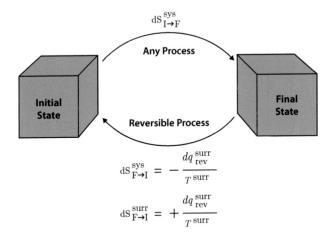

$$dS_{F \to I}^{sys} = - \frac{dq_{rev}^{surr}}{T^{surr}}$$

$$dS_{F \to I}^{surr} = + \frac{dq_{rev}^{surr}}{T^{surr}}$$

Figure 6.9: Thought experiment of entropy changes in a cyclic process

$$\Delta S^{surr} = \frac{q^{surr}}{T^{surr}} \tag{6.17}$$

$$dS^{surr} = \frac{dq^{surr}}{T^{surr}}. \tag{6.18}$$

We can construct a thought experiment to **determine the change in entropy of the system**. Assume we can go from an initial state I to a final state F with a corresponding change in the entropy of the system given by $dS_{I \to F}^{sys}$ (Figure 6.9). Assume further that you can return to the initial state by a **reversible** process by transferring heat from the surroundings. In that case, the change in entropy of the surroundings is, by definition, $\frac{dq_{rev}^{surr}}{T^{surr}}$ since the transfer of heat is done reversibly. The reversible return to the initial state changes the entropy of the system and the surrounding by the same, but opposite, amount so that $dS_{F \to I}^{sys} = - \frac{dq_{rev}^{surr}}{T^{surr}}$. Moreover, the overall change in entropy of the system is zero since it is a state function so

$$dS_{I \to F}^{sys} = -dS_{F \to I}^{sys}$$

$$= - \frac{dq_{rev}^{surr}}{T^{surr}}. \tag{6.19}$$

Since the process is reversible, the temperature of the system and the surroundings are the same at all times, hence $T^{surr} = T^{sys}$. Further, the heat transferred reversibly from the surroundings to the system are of equal value but of opposite signs, so $dq_{rev}^{surr} = -dq_{rev}^{sys}$. Combining these equalities, we finally get the system entropy change for the initial change as a function of the reversible heat transfer in the system and the temperature of the system

$$dS_{I \to F}^{sys} = \frac{dq_{rev}^{sys}}{T^{sys}}. \tag{6.20}$$

This can be integrated so that

$$\Delta S^{sys} = S_F - S_I = \int_I^F \frac{dq_{rev}^{sys}}{T^{sys}}. \tag{6.21}$$

We note that the entropy of the system increases, $\Delta S > 0$, whenever there is heat transferred to the system, $q^{sys} > 0$, which is consistent with the notion that the entropy is a measure of increased dispersal of energy (or increased disorder) in the system.

Note that the entropy is defined in terms of heat transferred to the surroundings. The entropy of the system is related to transfer of heat to or from the surroundings only during a reversible process.

Consider a closed system in thermal equilibrium with the surroundings (so that any process is isothermal). When the system changes reversibly, the change in entropy of the system and the surroundings must be zero. When the system changes irreversibly, or spontaneously, the overall entropy must increase, so we infer that

$$dS^{sys} + dS^{surr} \geq 0, \tag{6.22}$$

but the entropy of the surroundings changes by the amount of heat transferred **from** the system ($dq^{surr} = -dq^{sys}$) which leads us to the Clausius Inequality

$$dS^{sys} \geq \frac{dq^{sys}}{T^{sys}}. \tag{6.23}$$

This is significant since it relates the change in the entropy of the system to the transfer of heat from the system. In an isolated system there is no transfer of heat to or from the system, so $dS^{sys} \geq 0$ for any isolated system, where the greater than corresponds to the spontaneous (irreversible) process (Figures 6.6 and 6.7) and the equality corresponds to the reversible process. This leads directly to one of the formulations of the Second Law of Thermodynamics:

The entropy of an isolated system must increase during a spontaneous process.

One example of an isolated system is the Universe, so a consequence of the Second Law of Thermodynamics is that the entropy of the Universe must increase for all spontaneous processes. This is consistent with the notion that the system and the surroundings constitute the Universe.

The significance of the Second Law of Thermodynamics is that it provides a unique criterion for determining that a process is spontaneous. It is, however, not a practical criterion and we will explore alternatives in the next section.

6.2.3 The Third Law

The Third Law of Thermodynamics derives from the observation that there is a minimum absolute temperature, $T = 0$, which has never been attained. This leads to one expression of the Third Law:

> It is not possible to reach the absolute temperature of zero in a finite number of steps.

The temperature of absolute zero would have to correspond to the point at which all energy of any type of motion has been removed from the system; it is a solid in which there are no vibrations, no rotations, and there is perfect ordering of the molecules in a crystalline form. This leads to an alternative expression of the Third Law which links the concept of molecules to the concept of entropy and order[3]:

> "If the entropy of every **element** in its most stable state at T=0 is taken as zero, every **substance** has a positive entropy which at T=0 may become zero and does become zero for all perfect crystalline substances, including compounds."

The significance of the Third Law of Thermodynamics as stated here is that it provides a reference point for calculating absolute values of the entropy for any substance. This is in contrast to energy, which cannot be calculated in absolute terms.

6.3 GENERAL CRITERIA FOR SPONTANEOUS CHANGE

In order to arrive at more convenient criteria for deciding whether a process is spontaneous or not, we reexamine the Clausius Inequality for isochoric and isobaric processes that involve only PV-work where we can relate the heat transfer for the system to respectively the internal energy and the enthalpy.

For the isochoric process, we know that $dU = (dq^{sys})_V$. Hence the Clausius Inequality becomes

$$T^{sys} dS^{sys} \geq dU. \tag{6.24}$$

We can impose an additional constraint of either constant internal energy

---

[3] Atkins p. 115.

(as in an isolated system) or constant entropy (as in a reversible process) and we obtain[4]

$$(dS)_{V,U} \geq 0 \tag{6.25}$$
$$(dU)_{V,S} \leq 0. \tag{6.26}$$

The first inequality describes the criterion for a spontaneous process that the entropy must increase when the volume and internal energy are kept constant, which we saw earlier. The second inequality describes the new criterion that for a spontaneous process the internal energy must decrease when the volume and the entropy are kept constant.

Correspondingly, for the isobaric process we defined the enthalpy such that $dH = (dq^{sys})_P$, so now the Clausius Inequality is

$$T^{sys}dS^{sys} \geq dH \tag{6.27}$$

and thus for constant enthalpy processes and constant entropy processes we get respectively

$$(dS)_{P,H} \geq 0 \tag{6.28}$$
$$(dH)_{P,S} \leq 0. \tag{6.29}$$

The first inequality describes a new criterion for a spontaneous process that the entropy must increase when the pressure and enthalpy are kept constant. The second inequality describes another new criterion that for a spontaneous process the enthalpy must decrease when the pressure and the entropy are kept constant.

While we have new and different criteria for spontaneous processes, it is difficult to design experiments in which either the internal energy, the enthalpy, or the entropy is kept constant. It is experimentally much easier to design isochoric, isobaric, or isothermal processes, i.e., keeping volume, pressure, or temperature constant. To resolve this problem we **define** two new functions by subtracting the energy term TS[5] from respectively the internal energy and the enthalpy:

$$A = U - TS \tag{6.30}$$
$$G = H - TS. \tag{6.31}$$

The function A[6] is called the **Helmholtz Free Energy**. The function G is called the **Gibbs Free Energy**. The total differentials for these are respectively

---

[4]We remove the superscript that designates the system, since all the parameters relate to the system, so from now on, unless otherwise indicated, the parameters are those of the system.

[5]Note that TS also has units of energy.

[6]In some texts this is given the symbol F.

$$dA = dU - TdS - SdT \qquad (6.32)$$
$$dG = dH - TdS - SdT \qquad (6.33)$$

from which it is easy to show (see problem assignment)

$$(dA)_{T,V} \leq 0 \qquad (6.34)$$
$$(dG)_{T,P} \leq 0. \qquad (6.35)$$

The first inequality describes the criterion that for a spontaneous isothermal and isochoric process, the Helmholtz Free Energy must decrease. The second inequality describes the criterion that for a spontaneous isothermal and isobaric process, the Gibbs Free Energy must decrease. In all cases, we are constrained to systems that can only do PV-work.

The relations in Equations 6.34 and 6.35 are important since they refer to conditions that are relatively easy to control and it follows from these that a system will spontaneously move towards the condition where the (free) energy is minimal. The concept of energy minimization is well understood from mechanical systems and we see that represented in the thermodynamic concept as minimization of energy, described either as the minimum internal energy for isochoric and isentropic processes, as the minimum enthalpy for isobaric and isentropic processes, as the minimum Helmholtz Free Energy for isochoric and isothermal processes, and as the minimum Gibbs Free Energy for isobaric and isothermal processes.

The Gibbs Free Energy is most frequently used to determine whether a process is spontaneous or not because most processes we are concerned with experimentally are conducted in a closed system at constant pressure (atmospheric pressure) and constant temperature (room temperature). However, we will encounter examples where the Helmholtz Free Energy is the better function to use since the volume is constant.

It is interesting to note that surfaces, which are important for the properties of many nanoscale systems, are two-dimensional systems rather than three-dimensional systems. In that case, the surface area is substituted for the volume in the above relationships. In nanoscale systems where the surface areas remain constant, the Helmholtz Free Energy becomes the most relevant energy function as we shall see later.

6.4   THE MAXWELL RELATIONS

The First and Second Laws of Thermodynamics are framed in the context of isolated systems, but they can be combined to provide relationships that are general for closed systems in which only PV-work is possible.

From the First Law we write $dU = dq + dw$. For reversible processes, $dq = TdS$ and $dw = -PdV$ (only PV-work), so $dU = TdS - PdV$. But the internal energy is independent of the process, so this must hold for irreversible processes as well.

From the definitions of the enthalpy, $H = U + PV$, the Helmholtz Free Energy, $A = U - TS$, and the Gibbs Free Energy, $G = H - TS$ we can determine (see problem assignment) that any process in a closed system in the absence of non-PV-work will obey the following relationships

$$dU = TdS - PdV \tag{6.36}$$
$$dH = TdS + VdP \tag{6.37}$$
$$dA = -SdT - PdV \tag{6.38}$$
$$dG = -SdT + VdP. \tag{6.39}$$

Each of the energy functions are state functions, which are independent of the path. This also means that they are exact differentials for which the second derivatives are equal [reference Kreyszig p 59]. This leads us to the Maxwell Relations:

$$\left(\frac{\partial T}{\partial V}\right)_S = -\left(\frac{\partial P}{\partial S}\right)_V \tag{6.40}$$
$$\left(\frac{\partial T}{\partial P}\right)_S = \left(\frac{\partial V}{\partial S}\right)_P \tag{6.41}$$
$$\left(\frac{\partial P}{\partial T}\right)_V = \left(\frac{\partial S}{\partial V}\right)_T \tag{6.42}$$
$$\left(\frac{\partial V}{\partial T}\right)_P = -\left(\frac{\partial S}{\partial P}\right)_T. \tag{6.43}$$

A number of useful relationships can be derived from the Maxwell Relations and the exact differentials listed above. These are summarized in Table 6.1.

We have established that any of the energy functions can be used to determine whether a process is spontaneous or not (See Equations 6.26 6.29, 6.34, and 6.35) but each requires that a different pair of parameters are kept constant. For example, we can use the decrease in Gibbs Free Energy at constant temperature and pressure as a criterion for spontaneous change. However, if we want to determine whether the same process is spontaneous at a different temperature and pressure, we need to know how the Gibbs Free Energy changes with temperature and pressure, which can be established from Equation 6.39. We will use this concept further in Chapter 7 for example when we examine the temperature and pressure dependence of phase equilibria.

Table 6.2 interrelates some other useful quantities measurable by properly designed experiments.

6.5 OPEN SYSTEMS AND THE CHEMICAL POTENTIAL

The conclusions so far relate to isolated or closed systems in which the mass or the numbers of molecules cannot change. We describe the energy of such

$dU = TdS - PdV$	$T = \left(\frac{\partial U}{\partial S}\right)_V \quad \mid \quad C_V = \left(\frac{\partial U}{\partial T}\right)_V$
$\left(\frac{\partial T}{\partial V}\right)_S = -\left(\frac{\partial P}{\partial S}\right)_V$	$P = -\left(\frac{\partial U}{\partial V}\right)_S \mid dU = C_V dT - \left[P - T\left(\frac{\partial P}{\partial T}\right)_V\right]dV$
$dH = TdS + VdP$	$T = \left(\frac{\partial H}{\partial S}\right)_P \quad \mid \quad C_P = \left(\frac{\partial H}{\partial T}\right)_P$
$\left(\frac{\partial T}{\partial P}\right)_S = \left(\frac{\partial V}{\partial S}\right)_P$	$V = \left(\frac{\partial H}{\partial P}\right)_S \mid dH = C_P dT + \left[V - T\left(\frac{\partial V}{\partial T}\right)_P\right]dP$
$dA = -SdT - PdV$	$S = -\left(\frac{\partial A}{\partial T}\right)_V \mid$
$\left(\frac{\partial S}{\partial V}\right)_T = \left(\frac{\partial P}{\partial T}\right)_V$	$P = -\left(\frac{\partial A}{\partial V}\right)_T \mid dS = \frac{C_V}{T}dT + \left(\frac{\partial P}{\partial T}\right)_V dV$
$dG = -SdT + VdP$	$S = -\left(\frac{\partial G}{\partial T}\right)_P \mid$
$\left(\frac{\partial S}{\partial P}\right)_T = -\left(\frac{\partial V}{\partial T}\right)_P$	$V = \left(\frac{\partial G}{\partial P}\right)_T \mid dS = \frac{C_P}{T}dT + \left(\frac{\partial V}{\partial T}\right)_P dP$

systems at a particular temperature and pressure in terms of the Gibbs Free Energy, $G(T, P)$. If the temperature or the pressure changes, the change in the free energy is calculated by

$$dG = \left(\frac{\partial G}{\partial T}\right)_P dT + \left(\frac{\partial G}{\partial P}\right)_T dP = -SdT + VdP. \qquad (6.44)$$

In an open system the composition of the systems can change and this will affect the total energy of the system. If there is only a single component, the Gibbs Free Energy must now depend on the number of moles, n, of that component in the system as well, thus $G(T, P, n)$ and the change in free energy is now calculated as

$$dG = \left(\frac{\partial G}{\partial T}\right)_{P,n} dT + \left(\frac{\partial G}{\partial P}\right)_{T,n} dP + \left(\frac{\partial G}{\partial n}\right)_{T,P} dn$$
$$= -SdT + VdP + \left(\frac{\partial G}{\partial n}\right)_{T,P} dn$$
$$= -SdT + VdP + \mu dn. \qquad (6.45)$$

The last of these equations introduces the symbol μ to describe the Gibbs Free Energy per mole at constant temperature and pressure. The **molar Gibbs**

Table 6.2: Useful Measurable Quantities

Isobaric compressibility of thermal expansion:	$\alpha = \left(\frac{1}{V}\right)\left(\frac{\partial V}{\partial T}\right)_P)$
Isothermal compressibility:	$\kappa_T = -\left(\frac{1}{V}\right)\left(\frac{\partial V}{\partial P}\right)_T$
Isentropic compressibility:	$\kappa_S = -\left(\frac{1}{V}\right)\left(\frac{\partial V}{\partial P}\right)_S$
Joule–Thompson coefficient:	$\mu_{JT} = \left(\frac{\partial T}{\partial P}\right)_H$
and	$\mu_{JT}C_P = -\left(\frac{\partial H}{\partial P}\right)_T$
Combining these:	$C_P - C_V = \left(\frac{\partial\alpha^2}{\partial\kappa}\right)TV$

Free Energy is defined as the chemical potential of the component present in the system.

If there are many components, 1, 2, 3 etc., then there will be a chemical potential associated with each of these components since the change in free energy is now described by

$$dG = \left(\frac{\partial G}{\partial T}\right)_{P,n_1,n_2,n_3,\ldots} dT + \left(\frac{\partial G}{\partial P}\right)_{T,n_1,n_2,n_3,\ldots} dP + \left(\frac{\partial G}{\partial n_1}\right)_{T,P,n_2,n_3,\ldots} dn_1$$

$$+ \left(\frac{\partial G}{\partial n_2}\right)_{T,P,n_1,n_3,\ldots} dn_2 + \left(\frac{\partial G}{\partial n_3}\right)_{T,P,n_1,n_2,\ldots} dn_3 + \ldots$$

$$= -SdT + VdP + \mu_1 dn_1 + \mu_2 dn_2 + \mu_3 dn_3 + \ldots. \tag{6.46}$$

The chemical potential for component i is therefore

$$\mu_i = \left(\frac{\partial G}{\partial n_i}\right)_{T,P,n_{i\neq j}} dn_i. \tag{6.47}$$

The concept of chemical potential is extremely important since it allows us to deal with open systems. In addition, the chemical potential provides a singular approach to understanding spontaneous change; the system will always try to minimize the chemical potential of each of the components in the system. It is particularly noteworthy that the chemical potential of a component, in addition to being the molar Gibbs Free Energy at constant temperature and pressure, is also equal to the molar Helmholtz Energy at constant temperature and volume, the molar Enthalpy at constant entropy and pressure, and the molar internal energy at constant entropy and volume. i.e., the chemical potential for component i is

$$
\begin{aligned}
\mu_i &= \left(\frac{\partial G}{\partial n_i}\right)_{T,P,n_{i\neq j}} dn_i = \left(\frac{\partial A}{\partial n_i}\right)_{T,V,n_{i\neq j}} dn_i \\
&= \left(\frac{\partial H}{\partial n_i}\right)_{S,P,n_{i\neq j}} dn_i = \left(\frac{\partial U}{\partial n_i}\right)_{S,V,n_{i\neq j}} dn_i.
\end{aligned} \tag{6.48}
$$

The chemical potential is called a potential in analogy with a mechanical potential or an electrical potential and there is an associated chemical or thermodynamic force ($F_{thermodynamic} = -\nabla\mu$) which is able to do work. This is illustrated for example when an osmotic pressure is created from a chemical gradient across a semi-permeable membrane.

In the next chapters we will apply the chemical potential and the other thermodynamic functions to systems of interest in the field of nanoscience and nanotechnology.

CHAPTER 7

Surface thermodynamics

The previous chapter provided a review of the fundamentals of Thermodynamics to the point of introducing the chemical potential as the molar free energy. Moreover, we established the criteria for spontaneous processes to occur, namely minimizing the Gibbs Free Energy at constant temperature and pressure or the Helmholtz Free Energy at constant temperature and volume, or more generically, minimizing the chemical potential for each component in the system.

In this chapter, we will explore the consequences of applying these concepts to some general properties of nanoscale systems, and in particular begin to understand what the effect of the size of a particle is on particular system properties.

7.1   PHASE EQUILIBRIA

Let us consider a system of a single component at a constant temperature and pressure. The temperature and pressure dependence of the free energy is given by Equation 6.44. The temperature dependence of the molar free energy, or chemical potential, is therefore related to the molar entropy as suggested by Table 6.1:

$$\left(\frac{\partial G_m}{\partial T} \right)_P = \left(\frac{\partial \mu}{\partial T} \right)_P = -S_m, \tag{7.1}$$

where the subscript m indicates the molar quantity.

It is important to note that, because of the Third Law of Thermodynamics, the value of the molar entropy is positive at all temperatures greater than absolute zero. As a consequence, the chemical potential will always decrease as the temperature increases, which is in accord with the general notion that the system gets more disordered as the temperature increases.

We expect the entropy of a solid to be less than that of a liquid, which leads to the general observation that the rate of change of the chemical potential with temperature is less for a solid than for a liquid, i.e., the slope of a plot of

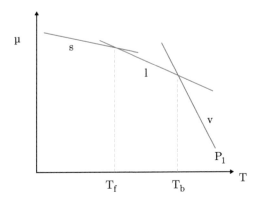

Figure 7.1: Schematic illustration of the variation of the chemical potential with temperature at constant pressure for a solid (s), a liquid (l) and a vapor (v).

the chemical potential as a function of temperature is less for the solid than for the liquid, as illustrated in Figure 7.1.

The temperature at which the two curves intersect is the temperature at which the chemical potential of the solid is identical to the chemical potential of the liquid, so the solid is in equilibrium with the liquid. This defines the solid-liquid phase transition temperature, T_f, at the constant pressure in question. For water, this would be $T_f = 273.16\ K$ at $P = 1\ bar$. If the solid is heated to a temperature greater than the phase transition temperature, the chemical potential of the solid is greater than that of the liquid, and the solid will spontaneously transform to the liquid, it will spontaneously melt. Correspondingly, if the liquid is cooled to a temperature lower than the phase transition temperature, the chemical potential of the liquid is greater than that of the solid and the liquid will spontaneously transform to the solid, it will spontaneously freeze (Figure 7.1).

Correspondingly, the entropy of the vapor is expected to be greater than that of the liquid, and hence the slope of the chemical potential of the vapor is even steeper than that of the liquid (see Figure 7.1). Again, there will be a point of intersection that defines the liquid-vapor phase transition temperature, T_b at the constant pressure in question. For water, this would be $T_b = 373.16\ K$ at $P = 1\ bar$. A liquid heated above the phase transition temperature will spontaneously vaporize while a vapor cooled below the phase transition temperature will spontaneously condense to the liquid.

Correspondingly, Equation 6.44 and Table 6.1 shows that the pressure dependence of the chemical potential at constant temperature is equal to the molar volume of the substance:

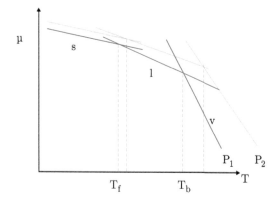

Figure 7.2: Schematic illustration of the variation of the chemical potential with temperature at two pressures for a solid (s), a liquid (l), and a vapor (v).

$$\left(\frac{\partial G_m}{\partial P}\right)_T = \left(\frac{\partial \mu}{\partial P}\right)_T = V_m. \tag{7.2}$$

Since the molar volume is always positive, the chemical potentials will always increase as the pressure increases. The net result is that the curves in Figure 7.1 move up as the pressure increases and move down as the pressure decreases. Normally, the molar volume of a solid is less than that of a liquid which suggests that the solid curve will move less than the liquid curve when the pressure changes (Figure 7.2). Similarly, the molar volume of the vapor is usually greater than that of the liquid, so the vapor curves will move more than the liquid curve. The net effect is that the transition temperatures (T_f and T_b) will increase whenever the pressure increases and will decrease whenever the pressure decreases and in each case the change is greater for the boiling than for the freezing temperature.

The exception is water, where the molar volume of ice is greater than that of water at the freezing temperature, which means that the freezing point of water decreases as pressure is increased. This turns out to be very important since it ensures that lakes and oceans don't freeze as the pressure increases at greater depths.

It is interesting to note that there is a combination of temperature and pressure at which all three curves will intersect. This corresponds to the **triple point** at which the chemical potentials of the solid, liquid, and vapor are all equal. At still lower temperatures and pressures, the chemical potential of the liquid is always greater than either that of the solid or of the vapor — the

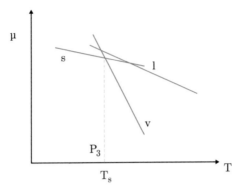

Figure 7.3: Schematic illustration of the variation of the chemical potential with temperature for a solid (s), a liquid (l), and a vapor (v) at pressures where the liquid is never stable leading to solid-vapor equilibria at the sublimation temperature.

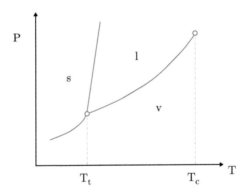

Figure 7.4: Schematic pressure-temperature phase diagram.

liquid is never stable — and the solid is in direct equilibrium with the vapor at the sublimation temperature T_s (Figure 7.3)

We normally plot the combination of temperatures and pressures at which the various phases are in equilibrium as a function of the temperature and pressure to obtain a phase diagram as illustrated in Figure 7.4.

Here the lines represent the temperatures and pressures where the chemical potential of adjacent phases are equal. The point where the three lines intersect is the triple point and the point where the liquid-vapor curve ends is the critical point beyond which it is impossible to tell the difference between the liquid and the vapor. As expected, the slope of the solid-liquid equilibrium

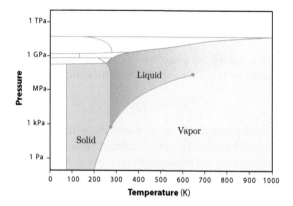

Figure 7.5: Schematic phase diagram of water with the pressure on a logarithmic scale

line is steeper than the solid-vapor and the liquid-vapor lines since a change in pressure will affect the vapor phase more than either the solid or the liquid.

As indicated above, the phase diagram for water is qualitatively different because the density of ice is less than that of water at the freezing point. This is why ice floats and why the lakes and the oceans do not freeze at the bottom. It also contributes to your ability to skate, since the pressure of the sharp edges of the skates melts the ice immediately below the skate, providing a thin film on which you can slide. The phase diagram for water is complex with a number of solid phases, as indicated in Figure 7.5.

The change in chemical potential for small, but finite changes in pressure at constant temperature can be estimated for solids and liquids by integrating Equation 7.2 assuming that the molar volume is almost independent of pressure to give

$$\Delta\mu = \mu_f - \mu_i = \int_{P_i}^{P_f} V_m dP = V_m(P_f - P_i) = V_m \Delta P. \qquad (7.3)$$

However, this does not work for the vapor since the molar volume of the vapor is very sensitive to pressure. As a first approximation, Equation 7.2 can be integrated assuming the vapor behaves ideally, so that $PV = nRT$ or $PV_m = RT$. Then

$$\Delta\mu = \mu_f - \mu_i = \int_{P_i}^{P_f} \frac{RT}{P} dP = RT ln \left(\frac{P_f}{P_i} \right). \qquad (7.4)$$

We will use the relationships in Equations 7.3 and 7.4 repeatedly in the following sections.

7.2 SURFACE TENSION

7.2.1 The Laplace equation

Consider a surface of a liquid. Expanding the surface requires work, ∂w, proportional to the change in the area, $d\sigma$. This can be expressed as

$$\partial w = \gamma d\sigma, \tag{7.5}$$

where the proportionality constant, γ, is called the **surface tension**. This is non-pressure volume work and is therefore related to the change in Helmholtz Free Energy, which is minimized whenever the surface area is minimized. The system will therefore always try to minimize the surface area.

For any given material, the minimal surface area is associated with a sphere of radius r, which has a surface area of $4\pi r^2$. The energy attributed to the surface must be $4\pi r^2 \gamma$.

Since the surface energy tries to minimize the surface area, then for a spherical object it can be viewed as creating a force acting inward along the radius of the sphere. The force is calculated as the negative of the gradient of the potential (here the surface energy) giving a value of $-8\pi r\gamma$, where the negative sign indicates a force acting towards the center of the sphere. The radial force per unit area, which is the pressure arising from the surface tension, is then

$$P_{sphericalsurface} = \frac{8\pi r\gamma}{4\pi r^2} = \frac{2\gamma}{r}. \tag{7.6}$$

This surface pressure will add to the external pressure acting on the sphere and when the sphere is at equilibrium the total external pressure must be balanced by the internal pressure of the material (the pressure arising from the lack of compressibility of the material) as illustrated in Figure 7.6. This leads to the Laplace equation:

$$P_{internal} = P_{external} + P_{sphericalsurface} = P_{external} + \frac{2\gamma}{r} \tag{7.7}$$

or

$$\Delta P = \frac{2\gamma}{r}. \tag{7.8}$$

The Laplace equation shows that as the radius of the sphere decreases, the internal pressure of the system must increase to maintain an equilibrium, which implies that smaller particles will be less stable and therefore harder to form or keep. As we will discuss later, forming a small, stable nanoparticle will require strong intermolecular adhesion forces.

7.2.2 The Kelvin equation

Let us now consider the effect of surface tension on a liquid drop in equilibrium with the vapor of the same material at a fixed temperature. The pressure of

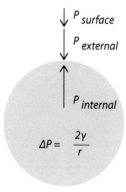

Figure 7.6: Illustration of the Laplace equation.

a vapor in equilibrium with a liquid is called the **vapor pressure** and is designated as P_r^* where the subscript refers to the radius of the drop. At equilibrium, the chemical potential of the liquid and the vapor are identical, and in the absence of any other pressures on the liquid, the pressure on the liquid is the vapor pressure. This would be true for a bulk material with a flat liquid surface exposed to the vapor.

When there is a change in pressure on the liquid, such as that arising from the surface tension on a curved surface, then the chemical potential of the liquid will increase according to Equation 7.3 where the molar volume is that of the pure liquid. However, to remain in equilibrium, the chemical potential of the vapor must change by the same amount. The change in chemical potential of the vapor will change according to Equation 7.4 assuming the vapor behaves ideally. We can then equate the two expressions for the change in chemical potential

$$\Delta\mu(l) = V_m(l)\Delta P = \Delta\mu(v) = RTln\left(\frac{P_r^*}{P_\infty^*}\right), \tag{7.9}$$

where we have assumed that the initial state is the liquid and vapor for a flat surface with $r = \infty$ and the final state is a spherical drop of radius r. Since the additional pressure on the liquid arises from the surface tension due to the curved surface, we can apply the Laplace equation in Equation 7.8 to get the Kelvin equation (see problem assignment)

$$P_r^* = P_\infty^* e^{\frac{2\gamma V_m(l)}{rRT}}. \tag{7.10}$$

The Kelvin equation demonstrates how *the vapor pressure of the liquid in a drop depends on the surface tension* of the material and on the radius of the drop. The vapor pressure of the liquid in a drop will always be greater than

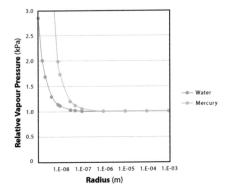

Figure 7.7: Variation of the relative vapor pressure of water and mercury with drop radius.

that of the liquid in the bulk with a flat surface. Moreover, as the radius of the drop decreases, the vapor pressure of the liquid in the drop increases.

Recall that a high vapor pressure corresponds to a liquid that is easily vaporized. Thus the material in a small drop is more readily vaporized than that in a large drop or in the bulk. This makes sense since the higher internal pressure would need to be overcome by stronger intermolecular forces. Moreover, as the surface curvature increases in the smaller drop there will be fewer intermolecular interactions from the nearest neighbors.

Figure 7.7 shows the variation of the vapor pressure of a liquid drop relative to the vapor of the bulk P_r^*/P_∞^* as a function of the drop radius for water and mercury. It is evident that the effect of the curvature is relatively small for both materials for drops larger than about 100 nm, but for drops smaller than that — nanodrops — the vapor pressure is quite sensitive to the drop radius.

This is manifest in a liquid such as water in an aerosol vaporizing faster that that in the bulk. This is also why very small drops of mercury are more hazardous than large drops of mercury. It should be noted, however, that the actual vapor pressure of mercury is less than 0.03% of that of water in the bulk and is less than 1% for 2 nm sized drops.

It is worth noting that the above discussion is equally valid for solid nanoparticles in equilibrium with their corresponding vapors during the process of sublimation. This would for example hold for ice particles (such as snow) at temperatures and pressures below the triple point.

The discussion so far has been focused on liquid drops in an atmosphere, for which the radius of curvature is positive. However, the same problem emerges when the radius of curvature is negative, that is, when the vapor is trapped as a bubble within a liquid phase. In that case, the surface tension will aim to minimize the surface area by decreasing the size of the bubble,

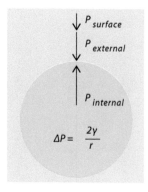

Figure 7.8: The pressures in bubbles.

again increasing the pressure of the vapor inside the bubble (Figure 7.8). In this case, the Kelvin equation is transformed by inverting the exponential and substituting the (positive) radius of the bubble for the (negative) radius of the liquid as shown in Equation 7.11

$$P_r^* = P_\infty^* e^{-\frac{2\gamma V_m\,(l)}{rRT}}.\tag{7.11}$$

Small bubbles of vapor are therefore inherently unstable, and in the limit, it requires a very high vapor pressure to form the first tiny bubble with just a few molecules in it. This can only happen if the temperature is increased significantly above the phase transition temperature and leads to super heating of the liquid and a sudden violent boiling of the liquid. However, in many cases there are small imperfections in the walls of the container that can form sites of nucleation of the bubbles. Alternatively, there are small impurities in the liquid (such as dust) which allows bubbles to form around them. In most laboratory settings we add highly porous materials (boiling chips or wood fibers) to act as nucleation sites to avoid super heating.

We can use the Kelvin equation to estimate the minimum size of a drop or a bubble that can be stable at a particular pressure $P > P^*$.

$$r_c = \frac{2\gamma V_m}{RT ln(P/P^*)}.\tag{7.12}$$

This is known as the critical radius of the object. Below the critical radius the nano-scale drop is unstable and will disappear by evaporation. Above the critical radius, the drop remains stable.

The treatment of the equilibrium between a liquid and its vapor can in principle be generalized to any transformation of a substance, S from one form to another

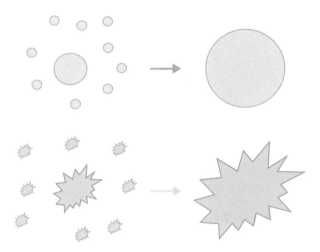

Figure 7.9: Schematic of the process of Ostwald ripening

$$S(l) \rightleftharpoons S(g)$$
$$S(s) \rightleftharpoons S(g)$$
$$S(s) \rightleftharpoons S(l) \tag{7.13}$$
$$S(l) \rightleftharpoons S(solution)$$
$$S(s) \rightleftharpoons S(solution).$$

The solid-vapor equilibrium would correspond to substances that can sublime, such as dry ice (solid carbon dioxide) while the solid-liquid equilibrium refers to the melting (or freezing) of the substance. In either case, the smaller the solid particle, the less stable it would be. The phenomenon of super-cooling of a liquid is a reflection of the fact that it is unfavorable for small solid particles to form, and frequently they will need nucleation sites to start the crystallization event.

The liquid-solution and solid-solution equilibria can be treated analogously if we substitute solubility S^* for the vapor pressure P^* in Equation 7.10 and the equivalent of the Kelvin equation is called the Ostwald–Freunlich relationship

$$S_r^* = S_\infty^* e^{\frac{2\gamma V_m(l)}{rRT}}. \tag{7.14}$$

This shows that the **solubility of a nanoparticle increases as the size of the particle decreases**. Analogously, we can estimate the critical size of a nanoparticle in a supersaturated solution by the equivalent of Equation 7.12.

The differential solubility as the size changes leads to the tendency for small objects to dissolve and to reform as large particles. This phenomenon is known as Ostwald ripening, as illustrated in Figure 7.9. Commonly cited examples of Ostwald ripening are the tendency for smooth ice cream to recrystallize while sitting for a long time in the fridge where small water crystals disappear to form larger crystals and therefore leading to a coarse, less palatable ice cream and for tiny droplets in suspensions to disappear and to form larger drops.

7.2.3 A molecular view of surface tension

Surface tension arises from the difference in intermolecular interactions in a liquid and a vapor. In fact, the liquid forms because the energy gained from favorable intermolecular interactions exceed the energy lost by the decrease in entropy in the liquid phase. This can be expressed thermodynamically from Equation 6.33 which at constant pressure and temperature shows us that

$$dG = dH - TdS \tag{7.15}$$

or for large changes

$$\Delta G = \Delta H - T\Delta S. \tag{7.16}$$

When the two phases are in equilibrium at the boiling or vaporization temperature T_{vap} the Gibbs Free Energy change must be zero, so

$$\Delta H_{vap} = T_{vap}\Delta S_{vap}, \tag{7.17}$$

where ΔH represents the heat needed to overcome the intermolecular interactions at constant pressure.

Since the intermolecular interactions are greater in the liquid, it implies that molecules at the surface will have fewer of these favorable intermolecular interactions, and hence they would prefer to transfer from the surface to the bulk. This is illustrated schematically in Figure 7.10. The net effect is that these favorable intermolecular interactions lead to an inward pressure and a tendency for the surface to shrink. This is therefore the source of surface tension. The stronger the intermolecular interactions, the greater the tendency to leave the surface, the larger the surface tension.

This effect can be illustrated by comparing the surface tension of molecules such as water and smaller alcohols as shown in Table 7.1. The surface tension of water is much higher than any of the alcohols because of the very strong hydrogen bonding. The surface tension rises in the series from methanol to 1-butanol suggesting increasing intermolecular interactions in the larger molecules. Correspondingly, the temperature at which they vaporize also increases.

7.2.4 Adhesion versus cohesion: Interfacial tension

The tendency of a liquid to minimize its surface area arises from the strong intermolecular interactions in the liquid relative to its vapor; this effect is

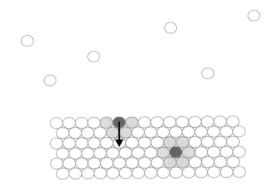

Figure 7.10: The molecular picture of surface tension

Table 7.1: Surface Tension and Boiling Point of Water and Some Alcohols

Compound	γ	T_{vap}
$H - OH$	$72.0\ 10^{-3}\ Nm^{-1}$	$373.1\ K$
$CH_3 - OH$	$21.5\ 10^{-3}\ Nm^{-1}$	$338.0\ K$
$CH_3CH_2 - OH$	$22.4\ 10^{-3}\ Nm^{-1}$	$351.5\ K$
$CH_3CH_2CH_2 - OH$	$24.2\ 10^{-3}\ Nm^{-1}$	$370.4\ K$
$CH_3CH_2CH_2CH_2 - OH$	$25.2\ 10^{-3}\ Nm^{-1}$	$390.3\ K$

termed **cohesion** of the liquid. However, the liquid will also form an interface with whatever container is holding it.

The interactions between liquid and the solid surface of the container may be stronger than the intermolecular interactions among the liquid molecules, in which case they will have a tendency to adhere to the surface. An example is the **adhesion** of water to a highly polar surface, such as glass. The water will tend to spread across such a surface in spite of the high surface tension of water, because the adhesive forces are greater than the cohesive forces. At some point, however, there will be an equilibrium since the energy needed to increase the surface area will balance the energy gained from the adhesion.

Conversely, the interactions between the liquid and the solid surface may be weaker than the intermolecular interactions in the liquid, in which case the liquid will be repelled from the surface and tend to form a nearly spherical

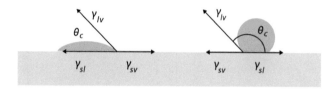

Figure 7.11: Illustration of the contact angle.

drop on top of the surface. The example of this would be water on a highly non-polar surface such as a plastic surface.

The interaction between two different materials A and B is characterized by the **interfacial tension** γ_{AB}

The equilibrium between the three regions, the solid (s), the liquid (l), and the vapor (v) would be characterized by three interfacial tensions: γ_{sl}, γ_{sv}, and γ_{lv} and it is possible to show (see problem assignment) that at equilibrium they are related by Young's equation

$$\gamma_{sl} = \gamma_{sv} - \gamma_{lv} cos\theta_c, \tag{7.18}$$

where θ_c is called the **contact angle** as illustrated in Figure 7.11. Note that when the contact angle is less than 90° the liquid adhesion is greater than the cohesion, while if it is greater than 90°, the liquid cohesion is greater than the adhesion.

If the liquid in the drop on the surface is water, then we consider the surface to be hydrophilic ("water loving") whenever the contact angle is less than 90°. This means that the adhesive forces of the water to the surface are greater than those among the water molecules in the drop. In contrast, we consider the surface to be hydrophobic ("water hating") whenever the contact angle is greater than 90° and in this case the adhesive forces between the water and the surface are less than those within the drop. Note that these terms relate to the macroscopic, thermodynamic properties of interfacial tensions, but they naturally arise from the molecular properties of the liquid and the surfaces. In other contexts, we refer to hydrophobic molecules as molecules that tend to avoid interaction with water at the molecular level. The hydrophobic interactions with water is usually interpreted as an entropic effect rather than an enthalpic effect since it is the disordering of the water structure

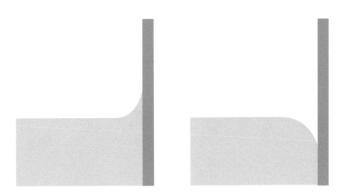

Figure 7.12: Illustration of adhesion or repulsion of a liquid at the solid-liquid-vapor interface.

around the hydrophobic molecule that is the major source of increased free energy.

Molecules that are hydrophobic in one part and hydrophilic in another part of the structure, are termed amphiphilic ("both loving"). Amphiphilic molecules will tend to self-assemble in water or organic solvents or to migrate to surfaces where they can significantly alter the interfacial tension. This is how soaps act to form micelles that can dissolve fatty materials.

When a liquid-vapor interface comes in contact with a solid surface at its edge, the liquid will either adhere to that solid surface (if $\theta_c < 90°$) and the meniscus will tend to curve up, or it will try to avoid contact with the surface (if $\theta_c > 90°$) and the meniscus will tend to curve down (Figure 7.12). This leads to the capillary effect which can result in liquids being sucked into fine pores or being repelled from them.

Consider a liquid with strong adhesive forces in contact with a vapor and air at a total pressure P and consider further a cylindrical tube inserted vertically into the liquid as illustrated in Figure 7.13. At the air-liquid interface outside the tube, the pressure on the surface is the total pressure P. At the air-liquid interface inside the tube, the pressure on the surface is also the total pressure, but because of the curvature of the liquid surface, the LaPlace equation tells us that there is an additional pressure upward, arising from the surface tension trying to minimize that surface area, which causes the liquid column to rise until the weight of the column of liquid is counteracted by gravity. The pressure arising from the surface tension must then equal the pressure arising from the liquid column rising to an height h, such that [see problem assignments]

$$h = \frac{2\gamma}{\rho g r}, \tag{7.19}$$

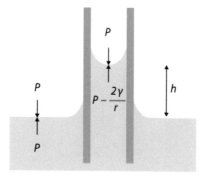

Figure 7.13: Illustration of capillary action when adhesion is greater than cohesion.

where ρ is the density of the liquid and g is the acceleration due to gravity.

Clearly, the smaller the radius of the cylinder, the higher the column can rise. For example for water with a density of 1 g cm^{-3} and a surface tension of 72 $10^{-3}N$ m^{-1} the height of the column of water for a 1 mm radius cylinder will be about 8 mm, while the height for a 1 micron radius cylinder will be about 8 m. In principle, therefore the capillary actions of nanoscale pores can be substantial in principle (see problem assignments), but on the other hand it might be unrealistic to have an 80 m tall cylinder of 100 nm radius. However, another way of looking at the problem, is that nanoscale pores in materials would be able to be filled quite easily with liquids through capillary action.

The situation where the liquid has low adhesion relative to the cohesion is illustrated in Figure 7.14. Now the surface tension causes the liquid column to lower within the cylinder, again by an amount calculable by Equation 7.19.

The curved liquid-air interface shown in Figures 7.13 and 7.14 is called the **meniscus** and that in Figure 7.13 is a concave meniscus while that in Figure 7.14 is a convex meniscus. These are commonly encountered in analog thermometers that use ethanol (with a concave meniscus) or mercury (with a convex meniscus). In either case, the pressure above the liquid is well below atmospheric pressure and they are closed systems so the position of the meniscus depend on the volume of the trapped air which in turn depends linearly on the temperature.

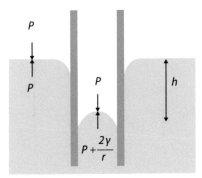

Figure 7.14: Illustration of capillary action when cohesion is greater than adhesion.

7.2.4.1 Superhydrophobicity

If a hydrophobic surface is structured, the contact angle increases to an extent that depends on the detailed structure and the nature of the interaction of the water with the structures. The effect of the surface structure has been explained by two models, the Wenzel model and the Cassie model.

The Wenzel model proposes that the increased contact angle arises from a larger area of contact as the surface area is increased from the smooth to the structured surface even when the projected area is the same. Thus the effective contact angle, θ^W is increased in proportion to the ratio, r, of the surface area of the structured surface to the surface area of the smooth surface and hence

$$cos\theta^W = rcos\theta, \qquad (7.20)$$

where $r > 1$.

The Cassie model proposes that the increased contact angle arises because air is trapped within the structures on the surface, and since the air-liquid contact angle will be close to $180°$, the contact angle will increase in proportion to the fraction of the drop in contact with the trapped air rather than the hydrophobic surface. If we term the fraction of the area where the surface is in contact with the drop f, then $1 - f$ is the fraction in contact with the air. Then

$$cos\theta^C = fcos\theta + (1 - f)cos\theta^{air} = fcos\theta - (1 - f), \qquad (7.21)$$

where $0 < f < 1$) and $\theta^{air} = 180°$.

The principles of these models are illustrated in Figure 7.15.

It has been argued that these models probably represents different limits of

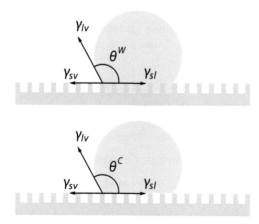

Figure 7.15: Illustration of the Wenzel and Cassie models of increased hydrophobicity on structured surfaces.

behavior, where the Wenzel model is most likely to hold for slightly hydrophobic surfaces i.e., when the contact angle is between $90°$ and some larger angle, θ^T, where there would be a transition to the behavior described by the Cassie model. In that case, the angle at which the transition in behavior occurs can be calculated by equating Equation 7.20 and 7.21 to give

$$cos\theta^T = \frac{f-1}{r-f},\qquad (7.22)$$

which yields a transition angle between $90°$ and $180°$ as expected.

When the surface contains nanostructures, the expectation is that either the ratio, r, is very large, of the fraction, f is very small. In either case, we expect to be beyond the transition angle and we would apply the Cassie model to give a very high contact angle. Such nanostructured surfaces are said to be superhydrophobic or ultrahydrophobic.

Interestingly, the mobility of the drop on the surface is expected to be quite different for the two models. The Wenzel model would predict that the drop would be pinned to the surface by the large contact area and it would require a high tilt angle for the drop to move across the surface. On the other hand, the Cassie model would predict that as the contact fraction, f, decreases, the drop should become more mobile on the surface. In fact, the mobility in a very highly hydrophobic surface can cause the drop to move across the surface and pick up small dirt particles, in effect causing a cleaning of the surface in the process. This self-cleaning effect of a drop on a highly structured surface commonly referred to as the lotus effect, since it is observed on the leaf of the lotus flower (as well as on other leaves). It is estimated that the structure of the lotus leaf reduces the fraction in contact to less than 0.01 or less than one percent.

The utility of ultrahydrophobic surfaces extends to commercial applications of coatings and paints and many other applications.

7.2.5 Examples

There are several examples that illustrate the principles discussed above and also demonstrate their limitations.

7.2.5.1 Properties of semiconductor nanoparticles

Farrell and Van Siclen argue in their work that there are limitations to the applicability of the Kelvin equation to solid nanoparticles of very small dimensions. They present several models that aim to link the thermodynamic arguments that lead to the surface tension concept to the intermolecular interactions among the atoms or molecules that make up the nanoparticles, building on the type of argument presented above in section 7.2.3.

They refer to several models that relate the binding energy E_b to the size of the particle (in their case they use the diameter, D, but for consistency we use the radius, r). In the Liquid Drop Model (LDM) the parameters that relate the binding energy to size are the volume of the atoms, v_0, and the surface tension, γ_n, where the subscript indicates that the surface tension may depend on the number of atoms in the particle when they are small

$$E_b = E_b^\infty \left(1 - \frac{3 v_0 \gamma_n}{r} \right). \tag{7.23}$$

In the Surface Areas Difference (SAD) model the parameters that relate the binding energy to the size of a spherical particle are the energies per unit area at the surface, γ_i, and in the crystal bulk, γ_0, and the crystallographic parameter $d_{hkl} = a/\sqrt{h^2 + k^2 + l^2}$ with a being the lattice parameter

$$E_b = E_b^\infty \left(1 - \frac{3 \left(\gamma_i / \gamma_0 \right) d_{hkl}}{2r} \right) \tag{7.24}$$

with a corresponding expression for the melting point T_m

$$T_m = T_m^\infty \left(1 - \frac{3 \left(\gamma_i / \gamma_0 \right) d_{hkl}}{2r} \right). \tag{7.25}$$

A corresponding expression for the melting behavior based on a thermodynamic model suggests that the parameters involved are the surface tensions of the crystal, γ_c, and the liquid, γ_l, as well as the latent heat of melting, C.

$$T_m = T_m^\infty \left(1 - \frac{3 \left(\gamma_c - \gamma_l \right)}{Cr} \right). \tag{7.26}$$

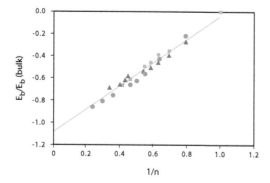

Figure 7.16: Illustration of Equation 7.27 for metals (Al, Au, Cs, W, Zr) [adapted from Farrel and Van Siclen].

In all cases, the prediction is an inverse linear dependence on the radius, which, for spherical particles is proportional to the cube root of the total number of atoms, N, i.e. $n = \sqrt[3]{N}$, and the radius of the atom in the particle, r_0, so that in general terms,

$$E_b^n = E_b^\infty \left(1 - \frac{c}{n}\right). \tag{7.27}$$

This general prediction was shown to hold for a number of metals, including aluminum (Al), gold (Au), cesium (Cs), tungsten (W), and Zirconium (Zr) as illustrated in Figure 7.16.

Farrel and Van Siclen then modeled group IV atoms (the non-metallic elements carbon (C), silicon (Si), and germanium (Ge) as well as the metallic elements tin (Sn) and lead (Pb)) and showed that in these systems the binding energy fits better with a inverse linear dependence on n^2 as expressed by Equation 7.28 and illustrated in Figures 7.17 and 7.18

$$E_b^n = E_b^\infty = \left(1 - \frac{c}{n^2}\right). \tag{7.28}$$

The proposed explanation for the greater dependence of the binding energy on the size of the nanoparticle is that the group IV elements have the ability to hybridize as the surface becomes more curved. In other words, while the number of nearest neighbors at the surface will decrease at smaller sizes, the strength of their interactions can be increased by forming additional bonds through $\pi - \pi$-interactions. These predictions suggest that several properties will behave differently for the group IV elements than expected for metallic elements; thus the melting point of smaller particles will be closer to the bulk value than expected for metallic particles of similar sizes; the vapor pressure will be closer to the bulk value; and the rate of reorganization through pro-

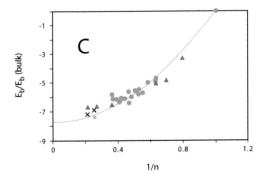

Figure 7.17: Illustration of binding energy dependence on n for various forms of carbon including nanotubes, buckyballs, and clusters [adapted from Farrell and Van Siclen].

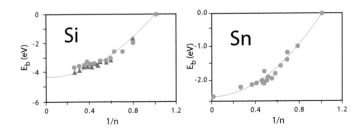

Figure 7.18: Illustration of binding energy dependence on n for Si and Sn [adapted from Farrell and Van Siclen].

cesses such as the Ostwald ripening will be slower; and the solubility will be less than otherwise expected.

7.2.5.2   Chemical equilibria are dependent on size

Recall that the chemical potential is the partial molar Gibbs Free Energy (Equation 6.47) and as a consequence we can determine the total Gibbs Free Energy, G as the sum of the product of the chemical potential, μ_i, and the number of moles, n_i, of each species, i, present in the system

$$G = \sum n_i \mu_i. \tag{7.29}$$

The total differential of this expression is then

$$dG = \sum n_i d\mu_i + \sum \mu_i dn_i. \tag{7.30}$$

But we also know from Equation 6.46 that at constant pressure and temperature

$$dG = \sum \mu_i dn_i. \tag{7.31}$$

Combining Equations 7.30 and 7.31 gives us the Gibbs–Duhem relations, namely that

$$\sum n_i d\mu_i = 0. \tag{7.32}$$

This shows us that the chemical potential of components in a mixture are dependent on each other.

A general chemical reaction can be expressed as

$$v_a A + v_b B + v_c C.... \rightleftharpoons v_m M + v_n N + v_p P...,$$

where the upper case letters refer to the reagent or product and v_x is the stoichiometric coefficient for reagent or product x. This can be generalized as

$$\sum_J v_J J = 0, \tag{7.33}$$

where $J = A, B, C, ...$ and v_J is negative for reagents and positive for products.

Solid phase components or solutes in this equilibrium are expressed in terms of their concentrations, c_i (or more generally their activities a_i) and all the vapor phase components are expressed in terms of their relative partial pressures, p_j/p_j^o, so the general reaction coefficient becomes

$$Q = \prod_i c_i^{v_i} \times \prod_j \left(\frac{P_j}{P_j^o} \right)^{v_j}. \tag{7.34}$$

The change in Gibbs Free energy associated with the reaction, $\Delta_r G$, is the sum of chemical potentials for each of the species in the reaction, μ_J, multiplied by their stoichiometric coefficients, v_J.

$$\Delta_r G = \sum_J v_J \mu_J$$
$$= \sum_J v_J \mu_J^o + RT ln Q \tag{7.35}$$
$$= \Delta_r G^o + RT ln Q.$$

Here the superscript ^{o} refers to the chemical potentials in the standard state, which is at pressures of 1 bar and concentrations of 1 M (or activities of 1).

When the reaction is at equilibrium, the change in total Gibbs Free energy is zero, so

$$\Delta_r G^0 + RT ln Q_{equilibrium} = 0. \tag{7.36}$$

The reaction coefficient becomes the equilibrium constant so, $Q_{equilibrium} = K$ and

$$\Delta_r G^o = -RT ln K. \tag{7.37}$$

The case where the reaction involves a combination of nanoparticles of radius, r, the argument has been made [Du et al.] that there are two components to the standard molar reaction Gibbs Free energy change — that from the bulk, $\Delta_r G_m^{bo}$ and that from the surface, $\Delta_r G_m^s$, which then yields

$$\Delta_r G_m^{bo} + \Delta_r G_m^s = -RT ln K. \tag{7.38}$$

The surface contribution from a nanoparticle made up of species i, of radius r_i, with surface tension γ_i, and molar volume V_{mi} can be estimated from the Kelvin equation (see Equation 7.9 and 7.10), where in this context the chemical potential associated with the surface for component i is[1]

$$\mu_i^s = \frac{2\gamma_i V_{mi}}{r_i}. \tag{7.39}$$

This leads to

$$\Delta_r G_m^s = \sum_i v_i \frac{2\gamma_i V_{mi}}{r_i} \tag{7.40}$$

and hence we obtain, using Equation 7.38, the effect of the size of the nanoparticles of relevant reactants and products on the molar Gibbs Free Energy of reaction

$$\Delta_r G_m = \Delta_r G_m^{bo} + \sum_i v_i \frac{2\gamma_i V_{mi}}{r_i}$$
$$= -RT ln K^b + \sum_i v_i \frac{2\gamma_i V_{mi}}{r_i} \tag{7.41}$$
$$= -RT ln K,$$

where the superscript ^{bo} again refers to the standard states in the bulk and

---

[1]In the reference from which this is drawn, the authors use the symbols σ_i for the surface tension and the ratio of the molar mass M_i and the density ρ_i rather than the molar volume such that $V_{mi} = M_i/\rho_i$.

where we have related the molar Gibbs Free Energy of reaction in the bulk standard state to the equilibrium constant of the bulk, K^b, through Equation 7.37.

Thus

$$ln\frac{K}{K^b} = -\frac{1}{RT}\sum_i v_i \frac{2\gamma_i V_{mi}}{r_i}.$$ (7.42)

It is evident that as the sizes of the nanoparticles decrease the influence of the surface tension becomes more pronounced. If the nanoparticle is a reactant, the stoichiometric coefficient is negative, so the smaller nanoparticle will cause the equilibrium constant to be larger than that in the bulk, shifting the reaction toward the product — *smaller reactant nanoparticles will shift the equilibrium to favor more product formation*. Conversely, *smaller product nanoparticles will shift the equilibrium to favor more reactant formation*.

The Gibbs–Helmholtz equation relates the temperature dependence of the Gibbs Free energy at constant pressure to the Enthalpy. From the definition of the Gibbs Free energy (Equation 6.31) and from its temperature dependence (Table 6.1) we can show [see problem assignments] that

$$\left(\frac{\partial (G/T)}{\partial T}\right)_P = -\frac{H}{T^2}.$$ (7.43)

Correspondingly, the molar enthalpy of reaction can be calculated as

$$\Delta_r H_m = -T^2 \left(\frac{\partial (\Delta_r G_m/T)}{\partial T}\right)_P,$$ (7.44)

which leads to an expression for the molar enthalpy of reaction when nanoparticles are involved

$$\Delta_r H_m = \Delta_r H_m^b + \sum_i v_i \frac{2V_{mi}}{r_i}\left[\gamma_i - T\left(\frac{\partial \gamma_i}{\partial T}\right)_P\right].$$ (7.45)

Finally, the molar entropy of reaction can be determined from the molar Gibbs Free energy of reaction by differentiation with respect to temperature at constant pressure

$$\Delta_r S_m = -\left(\frac{\partial \Delta_r G_m}{\partial T}\right)_P = \Delta_r S_m^b - \sum_i v_i \frac{2V_{mi}}{r_i}\left(\frac{\partial \gamma_i}{\partial T}\right)_P.$$ (7.46)

These relations were tested for the heterogeneous chemical reaction where nanoparticles of cuprous oxide $(CuO(nano))$ were reacted (dissolved) with aqueous sodium bisulfate $(NaHSO_4(aq))$ to give sodium sulfate $(Na_2SO_4(aq))$, copper sulfate $(CuSO_4(aq))$ and water (H_2O):

$$CuO(nano) + 2\ NaHSO_4(aq) \rightleftharpoons Na_2SO_4(aq) + CuSO_4(aq) + H_2O$$

and as shown in Figures 7.19 and 7.20 the prediction of a linear dependence on the inverse size of the nanoparticle is confirmed.

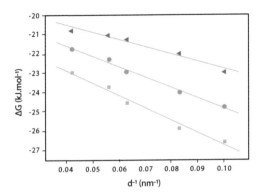

Figure 7.19: Dependence of the molar Gibbs Free energy of reaction on the size of nanoparticles [adapted from Du et al.].

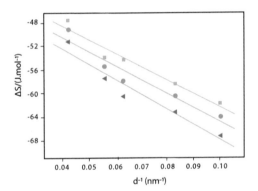

Figure 7.20: Dependence of the molar entropy energy of reaction on the size of nanoparticles [adapted from Du et al.].

7.2.5.3 Catalysis

The catalytic activity of a solid material depends on the activity of the surface material that can be in contact with the reactants, whether in solution or in a vapor phase. We expect that that there can be a number of contributing factors:

1. the increased number of surface atoms or molecules available as the surface to volume ratio changes;

2. the abilty of the surface atoms to bind to the reactants as surface geometries change in smaller particles;

Table 7.2: The Percent Surface Atoms as a Function of Nanoparticle Size

Diameter of particle	Number of Atoms in particle	% of atoms on the surface
10 nm	50,000	10
8 nm	22,700	14
6 nm	9,590	19
4 nm	2,840	28
2 nm	350	50
1 nm		80

3. the changes in the electronic properties of the surface states as the particle size decreases.

With respect to the first point, we note that the fraction of atoms at the surface increases dramatically as the particle gets very small (Table 7.2 from Binns *Introduction to Nanoscience and Nanotechnology* Figure 1.6).

With respect to the second point, we saw in the previous sections that the binding energy depends on the size of the particle, and that the equilibrium of the chemical reactions can be shifted in manners that depend on the particle size. In addition, the geometry of the molecules on a highly curved surface can expose more reactive coordination sites. This effect was illustrated by Lopez and coworkers who studied the catalytic activity of gold nanoparticles on the oxidation of carbon monoxide (CO) at low temperatures. In this work they analyzed the rate of oxidation of CO by gold nanoparticles supported on a number of solid surfaces as illustrated in Figure 7.21 and estimated that the rate depended on the inverse cube of the diameter of the gold nanoparticle, a significantly greater size dependence that expected from surface tension effects alone. The argument is that the d^{-3} dependence can arise if it is the free corners that contribute to the activity of the catalysts since this is where the reagents and products can bind to the low coordinate sites of the gold, which supports the third point.

The calculations of the binding of carbon monoxide and oxygen to different facets of a gold nanocrystal is shown in Figure 7.22 at the top and demonstrates that the binding is much stronger to the smaller nanocluster (Au_{10}) than to the flat surfaces such as the $Au(111)$ crystalline face. In fact, the authors estimate that this translates to a factor of 10^{13} in the relative binding of CO to the gold nanocluster compared to the flat surface.

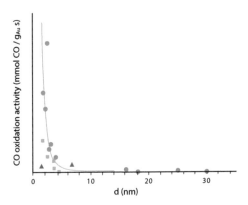

Figure 7.21: Dependence of oxidation of carbon monoxide on the size of gold nanoparticles supported on a variety of substrates [adapted from Lopez et al.].

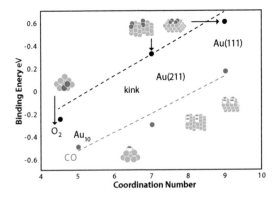

Figure 7.22: Relationship between energy of binding and the ability to for oxygen and carbon monoxide to coordinate with the gold depending on the surface structure of the gold nanoparticle [adapted from Lopez et al.].

CHAPTER 8

The work function

8.1 THE PHOTOELECTRIC EFFECT

The photoelectric effect, which was first explained by Einstein in 1905 and earned him a Nobel Prize, refers to the observations that photons of sufficient energy can cause electrons to be removed from a metal and cause an electric current to flow. The minimum energy needed to achieve this is the photoelectric work function, Φ. It can be determined by measuring the excess kinetic energy of the electrons emitted from a surface exposed to light of energy $h\nu$ such that

$$\Phi = h\nu - \frac{1}{2}m_e v^2. \tag{8.1}$$

The work function associated with a metal can be thought of as the energy needed to move an electron from the Fermi Level (which for a metal is the same as the Fermi Energy, see Section 5.1) to the vacuum without any residual kinetic energy as illustrated in Figure 8.1.

The work function associated with a semiconductor then is the energy needed to move an electron from the Fermi Level (which now represents the energy level where the Fermi–Dirac distribution[1] predicts a 50 % probability of finding the electron, see Equation 5.11) as illustrated in Figure 8.1.

There are a number of tools that can be used to measure the work function. The tutorials provided by Dr. Rudy Schlaf at the University of South Florida are well known and describe two approaches — the Kelvin Probe and Photoemission Spectroscopy. [2] We will focus on their description of Photoemission Spectroscopy (PES) as a useful illustration of the approaches.

---

[1] Recall that there are three important distributions of energy that governs the probability of particles having particular energies: the **Maxwell–Boltzmann Distribution** that pertains to indistinguishable particles such as ideal gas molecules; the **Bose–Einstein Distribution** that pertains to bosons that are particles with integer spins (n)such as phonons or Helium; and the **Fermi–Dirac Distribution** that pertains to fermions which are particles with half-integer spins (1/2 n) such as electrons or protons.

[2] (http://rsl.eng.usf.edu/Pages/Tutorials/TutorialsWorkFunction.pdf).

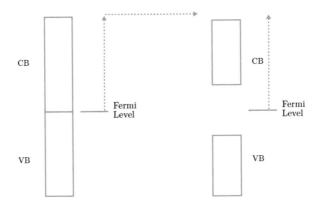

Figure 8.1: Illustration of the work function for a metal (left) and a semiconductor (right). VB stands for the valence band, CB stands for the conduction band. The dotted line illustrates the path of the electron to the vacuum level.

In this experiment, the kinetic energy of electrons emerging from a surface is measured as a function of the energy deposited on the surface using light. The kinetic energy E_{kin} will depend on the energy of the electron within the valence band — the binding energy of the electron, E_B — and the amount of energy deposited on the sample $h\nu$:

$$E_{kin} = \frac{1}{2}m_e v^2 = h\nu - E_b. \tag{8.2}$$

Figure 8.2 shows an example of a photoelectron spectrum in which the photon counts measured at a particular kinetic energy is plotted as a function of the binding energy. In this spectrum the Fermi Energy E_F corresponds to zero binding energy and the features at higher binding energy reflect the density of states within the valence band. The feature corresponding to inelastically scattered electrons at low energies arises from electrons excited within the material at a depth where they have to travel physically through the material before they reach the surface. During that process, they can scatter off the nuclei in the material and lose energy in the process. The secondary cut-off at E_C corresponds to the binding energy where the kinetic energy is zero. The work function can be calculated as the difference between the energy of the photons and the binding energy at that secondary edge.

The concept is further illustrated schematically in Figure 8.3. In this case the measured spectrum is shown relative to the kinetic energy at the top, this scale being the reverse of the binding energy curve seen in Figure 8.2. The lower part of the figure shows the expected density of state distribution of the valence band electrons as inferred by the shape of the photoelectron spectrum.

We expect, based on the discussion in Chapters 2-5, that the details of the photoelectron spectrum will depend on the density of states in the valence

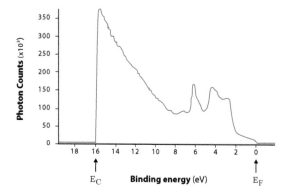

Figure 8.2: Example of a photoelectron spectrum in terms of photon counts as a function of binding energy [adapted from Schlaf tutorial].

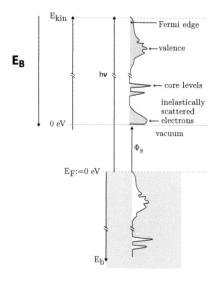

Figure 8.3: Illustration of the relationship between the photoelectron spectrum, the density of state of the valence band electrons, and the work function [adapted from Schlaf tutorial].

bands, which in turn would depend on the dimensions of the material and its lattice structure.

It has been demonstrated experimentally and theoretically by Zhou and Zachariah that the work function will be sensitive to the curvature of the material. In fact, the variation in the surface potential has been shown to depend

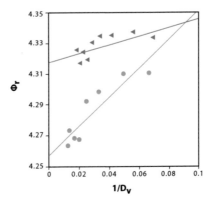

Figure 8.4: Variation of work function with particle diameter for silver nanoparticles [adapted from Zhou and Zachariah].

on the inverse of the cube root of the number of atoms in a spherical nanoparticle (or on the inverse of the volume equivalent diameter D_V) as illustrated in Figure 8.4, which compares the work function of spherical particles with those of aggregates of these particles.

$$\Phi_r = \Phi_\infty + \frac{a}{\sqrt[3]{N}} = \Phi_\infty + \frac{a}{D_V}. \tag{8.3}$$

CHAPTER 9

Understanding interactions between surfaces

The interaction in solution between small particles has been studied for more than a century within the field of colloid science. In this chapter we will look at the classical approaches to colloid science, which was primarily concerned with micron scale objects, and extend these concepts to the nanoscale.

9.1   BASIC TERMINOLOGY

The field of colloid science is filled with ambiguous terminology depending on the authors and the problem. In the context of this text, we will use the following descriptors:

1. Dispersion — the process whereby colloid particles suspend into individual units in a solvent, usually water.

2. Aggregation — the process whereby colloid particles associate to form larger structures that are *tightly* bound together and hence it is usually irreversibly. This is also termed coagulation in some contexts.

3. Flocculation — the process whereby colloid particles associate to form larger structures that are *loosely* bound and hence the association is usually reversible.

4. Coalescence — the process whereby colloid particles associate and *fuse* to form larger particles, perhaps by Ostwald ripening or by simple fusion.

5. Sedimentation — the process whereby aggregates or flocculates settle from suspension due to gravity, which can be enhanced by centrifugation.

6. Stabilization — the process whereby aggregation or flocculation is inhibited by agents that modify the surface properties of the colloid particles.

7. Brownian motion — the random movement of colloid particles in suspension due to thermal fluctuations and collisions with solvent molecules.

The behavior of colloid particles (or nanoparticles) in suspension depends on the balance of attractive and repulsive forces that act between their surfaces.

$$U = U_{attractive} + U_{repulsive} \qquad (9.1)$$

There are a number of attractive and repulsive interparticle forces that lead either to dispersion, flocculation, or aggregation depending on the details of the balance between them.

9.2 FORCES BETWEEN SURFACES

The forces between molecules or particles arise from a number of sources, some of which act over longer distances and others that act at short distances. These are described in detail in Chapter 29 (and Table 29.1) in Part D of the *Springer Handbook of Nanotechnology* authored by M. Ruth and J.N. Israelachvili.

The forces can arise because of quantum mechanical effects, such as formation of bonds; from electrostatic interactions between charged species, such as ions on the surface of the particles; from the interactions between transient or permanent dipoles on the surfaces; and from effects of the solvent in which the particles are dispersed.

1. Quantum mechanically based interactions: these are very short range interactions ($<$ nm) that lead to interactions of the electrons in the molecules on a surface. This can lead to very strong attractive forces through formation of bonds, or to very strong repulsive forces, often modeled as hard-core or steric interactions. They become very important at the molecular level, but are less important for interactions at longer range ($>$ nm).

2. Electrostatic interactions: these are long range interactions arising from the forces between two charges, q_1 and q_2 separated by a distance D_{12} as described by the Coulomb equation

$$F_{Coulomb} = \frac{1}{4\pi\epsilon_0}\frac{q_1 q_2}{D_{12}^2}, \qquad (9.2)$$

where ϵ_0 is the permittivity of free space. Since the force varies as the inverse separation squared, the potential will vary as the inverse of the separation. The force will be attractive (< 0) if the charges are of opposite signs, and repulsive (> 0) if they are of the same sign.

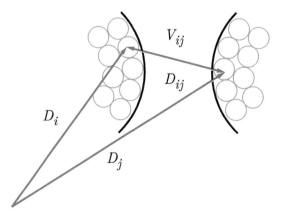

Figure 9.1: Illustration of the principle of summation of interactions between atoms in particles as in Equation 9.3.

3. Van der Waals interactions: these are attractive forces that arise from interactions between permanent or transient dipoles in the system. There are three components: those between sets of permanent dipoles (also referred to as Keesom interactions); those between permanent dipoles and induced dipoles (also referred to as Debye interactions); and those between transient and induced dipoles (also referred to as London dispersion interactions). While these interactions were developed for interactions between molecules — primarily in a vacuum — they can be extended to interactions between assemblies of atoms or molecules in particles by pairwise summation of the individual interactions (making the approximation that they act independently). This concept was developed by Hamaker and is expressed as

$$V_{total} = \frac{1}{2} \sum_{i=0}^{N} \sum_{j=0(\neq i)}^{N} V_{ij}(D_{ij}), \tag{9.3}$$

where V_{ij} is the potential between the $i'th$ and $j'th$ atom (or molecule) and D_{ij} is the separation between them as illustrated in Figure 9.1.

4. Solvation effects: these are modulations of the interactions between the surfaces that arise from the properties of the intervening material. These can be continuous properties, such as the refractive index and the dielectric constants of the solvent or properties that depend on the structure of the solvent and its ability to form specific interactions such as hydrogen bonds in water.

9.3 ATTRACTIVE FORCES

In solutions, the attractive forces between colloid particles or nanoparticles are dominated by the Van der Waals forces. Between individual molecules and atoms, these attractive forces depend on the inverse seventh power of the separation, which means that the potential between them depends on the inverse sixth power of the separation.[1] Using the Hamaker formulation for the summation of these potentials, one can determine the attractive interactions between surfaces of varying geometry (see Leckband and Israelachvili Quart. Rev. Biophys. 34, 105-267 (2001)). In general, these will depend on the geometry of the objects and their separation, D, such that

$$U_{attractive} = C_{attractive} D^{-m}, \tag{9.4}$$

where the exponent, m, of the separation, D, has values of 1, 1.5, or 2 depending on the geometry. The attractive interaction potentials are illustrated for several geometries in Table 9.1.

The constant, A_H, in Table 9.1 is the Hamaker constant. This constant is the sum of two major components: the first, $A_{H_{v=0}}$, arises from the interactions that involve a permanent dipole (the sum of the Keesom (dipole-dipole) and Debye (dipole-induced dipole) interactions) which depend on the dielectric constant (or permittivity), ϵ, in the interacting particles and of the material in the medium between them; the second, $A_{H_{v>0}}$, arises from the transient dipole interactions (the London dispersion interactions) which depend on the refractive index, n, of these materials. These properties are shown for two spheres in Figure 9.2.

$$A_H = A_{H_{v=0}} + A_{H_{v>0}}, \tag{9.5}$$

where

$$A_{H_{v=0}} = \frac{3}{4} kT \left(\frac{\epsilon_1 - \epsilon_3}{\epsilon_1 + \epsilon_3} \right) \left(\frac{\epsilon_2 - \epsilon_3}{\epsilon_2 + \epsilon_3} \right) \tag{9.6}$$

and

$$A_{H_{v>0}} = \frac{3h\nu}{8\sqrt{2}} \frac{\left(n_1^2 - n_3^2\right)\left(n_2^2 - n_3^2\right)}{\sqrt{(n_1^2 + n_3^2)}\sqrt{(n_2^2 + n_3^2)}\left[\sqrt{(n_1^2 + n_3^2)} + \sqrt{(n_2^2 + n_3^2)}\right]^{1/2}} \tag{9.7}$$

The first term is on the order of 3×10^{-21} J and is fairly insensitive to the nature of the materials, particularly if the solvent is water, where $\epsilon_3 = 80$, is much larger than that of most materials ($\epsilon_1 \approx 2 - 6$), so the ratios are close to unity. If the particles are made of metals, such as silver or gold, this term is nearly zero since there are no permanent dipoles on their surfaces.[2]

---

[1]This corresponds to the D^{-6} component of the empirical Lennard–Jones Potential, or the 6-12 potential where the D^{-12} component represents the repulsive potential arising from direct orbital interactions.

[2]Note that $kT \approx 4 \times 10^{-21}$ J at room temperature, which is approximately twice the

Table 9.1: Attractive Forces and Potentials between Surfaces of Different Geometry

Geometry	Force	$U_{attractive}$
Two flat surfaces	$\dfrac{-A_H}{6\pi D^3}$	$\dfrac{-A_H}{12\pi D^2}$
A sphere of radius r near a flat surface	$\dfrac{-A_H r}{6D^2}$	$\dfrac{-A_H r}{6D}$
A rod of radius r near a flat surface	$\dfrac{-A_H\sqrt{r}}{8\sqrt{2}D^{5/2}}$	$\dfrac{-A_H\sqrt{r}}{12\sqrt{2}D^{3/2}}$
Two spheres of radii r_1 and r_2	$\dfrac{-A_H}{6D^2}\left[\dfrac{r_1 r_2}{r_1+r_2}\right]$	$\dfrac{-A_H}{6D}\left[\dfrac{r_1 r_2}{r_1+r_2}\right]$
Two parallel rods of radii r_1 and r_2	$\dfrac{-A_H}{8\sqrt{2}D^{5/2}}\sqrt{\dfrac{r_1 r_2}{r_1+r_2}}$	$\dfrac{-A_H}{12\sqrt{2}D^{3/2}}\sqrt{\dfrac{r_1 r_2}{r_1+r_2}}$
Two perpendicular rods of radii r_1 and r_2	$\dfrac{-A_H\sqrt{r_1 r_2}}{6D^2}$	$\dfrac{-A_H\sqrt{r_1 r_2}}{6D}$

The second term ranges from values around 10^{-21} to 10^{-19} J, the larger values arising for metals such as silver and gold where the refractive indexes are an order of magnitude smaller than that of solvents such as water. Some examples of the values of the Hamaker constants for different materials suspended in a medium are shown in Table 9.2.

The variation in the attractive potential as a function of separation of two identical spheres is shown in Figure 9.3 for three systems: 100 nm radius liposomes, 10 nm radius protein structures, and 10 nm radius gold nanoparticles. Clearly, the second term of the Hamaker constant dominates the interactions for the gold nanoparticle which has a much stronger attractive potential for a given separation.

energy available per degree of freedom of motion in a molecule, for example, a monatomic ideal gas has the kinetic energy $\frac{3}{2}kT$ or $\frac{1}{2}kT$ per coordinate of movement. Correspondingly, the energy of a diatomic molecule in the gas phase is on the order $\frac{7}{2}kT$ or $\frac{3}{2}kT$ for the three translational motions, $\frac{3}{2}kT$ for the three rotational motions, and $\frac{1}{2}kT$ for the single vibrational motion. As a consequence, 10^{-21} to 10^{-20} J is therefore the order of magnitude of energy that can be imparted by a solvent molecule on a nanoparticle to cause it to move.

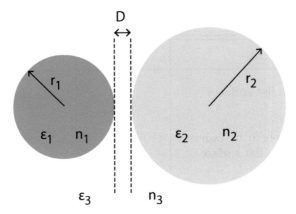

Figure 9.2: Illustration of parameters involved in calculating the attractive potential as per Equations 9.6 and 9.7 and the relations in Table 9.2.

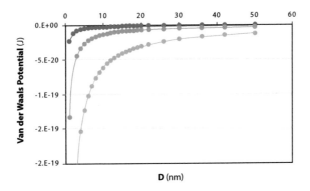

Figure 9.3: The attractive Van der Waals potential for two identical spheres as a function of separation — for 100-nm-radius liposomes (blue), for 10-nm-radius proteins (red), and for 10-nm-radius gold nanoparticles (gray).

It is interesting to note that the surface energy (or the surface tension), γ, is defined as half the absolute value of the interaction energy between two flat surfaces at a distance of contact, D_0. The surface tension can therefore be estimated from the first term in Table 9.1 as

$$\gamma = \frac{1}{2}\frac{A_H}{12\pi D_0^2}.$$

Table 9.2: Hamaker Constants for a Variety of Systems

Particle 1	Medium	Particle 2	A_H $(10^{-20}$ $J)$
Water	Hydrocarbon	Water	0.47
Hydrocarbon	Water	Hydrocarbon	0.47
Membrane	Water	Membrane	≈ 0.8
Polystyrene	Water	Polystyrene	1.4
Protein	Water	Protein	$1 - 1.5$
Air	Water	Air	3.7
Silver	Water	Silver	37
Gold	Water	Gold	34

The challenge is to know what the contact distance is. One expects it to be at the atomic dimension, or about 0.3 ± 0.15 nm. A value of 0.165 nm appears to be common for macroscopic surfaces as indicated here. For two hydrocarbon surfaces in contact through air, where $A_H = 5 \times 10^{-21}$ J, one finds $\gamma = \frac{5 \times 10^{-21}}{24\pi(1.65 \times 10^{-10})^2} = 24$ mN m^{-1}, which compares very well with the measured surface tension of about 25 mN m^{-1} at room temperature.

While the Van der Waals attractive forces are dominant for most systems, there are other attractive forces that can contribute and in some cases actually dominate. These include the forces arising from hydrophobic interactions between two hydrophobic surfaces; ionic forces between particles of opposite surface charge, and forces associated with hydrogen bonding which form permanent dipoles and are highly directional. These have been reviewed in detail, in the context of interactions in biological systems, by Leckband and Israelachvili (2001).

The Van der Waals potential assumes the solvent is homogeneous and therefore has no structure. In a more general treatment (see Leckband and Israelachvili), the attractive forces can be modeled as an oscillatory modulation of the exponentially decaying potential to account for layers of solvent between the surfaces where there are regions of space , corresponding to integral numbers of solvent molecular layers, where the attractive forces are maximal, and regions of space, corresponding to non-integral numbers of solvent molec-

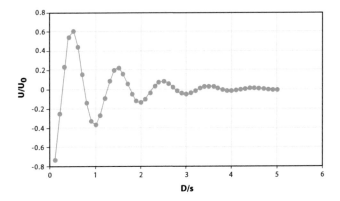

Figure 9.4: The oscillatory attractive potential as a function of separation in units of solvent dimension.

ular layers, where the attractive forces are minimal. As a consequence, there can be multiple separations at which there are stable interactions between the nanoparticles. Figure 9.4 illustrates this principle schematically in arbitrary energy units, U/U_0, and as a function of separation, D, in units of a parameter, s, related to the dimension of the solvent.

9.4 REPULSIVE FORCES BETWEEN CHARGED SURFACES

There are two critical repulsive forces at play in the interaction between particles in solution. The first is the short range contact repulsion, which can be modeled as hard spheres where the interaction energy goes to infinity when the particles make contact, or as a steric repulsion such as the 10^{-12} term in the Lennard–Jones potential. The second is a long range electrostatic repulsion at play between particles of similar charge.

To understand the origin of the electrostatic repulsion, we need to understand how a charged surface gives rise to a potential on the surface, which in turn generates a potential into the solution that can be affected by the properties of the solution, such as the concentration of ions in solution and their distribution near the surfaces.

The combination of the repulsive electrostatic forces and the attractive Van der Waals forces is often referred to the DLVO Theory (named for Derjaguin, Landau, Verwey, and Overbeek who developed the principles in the 1940s).

9.4.1 Electric potentials and surface potentials

The gradient or the divergence of the electric field, E, at some point in space is proportional to the density of charge, ρ, at that point. The proportionality constant is the inverse of the permittivity[3] of the medium in which the charge is positioned:

$$\nabla E = \frac{\rho}{\epsilon_r \epsilon_0}. \tag{9.8}$$

Further, the electric field is the gradient of the electric potential, ψ, so that $E = -\nabla \psi$ and if that is substituted into Equation 9.8 we get the **Poisson's equation**[4]

$$\nabla^2 \psi(r) = -\frac{\rho(r)}{\epsilon_r \epsilon_0}. \tag{9.9}$$

This relates the potential at any point in space to the charge density at that point in space. In the special case where there is a region of space with no charge density, $\rho(r) = 0$, the Poisson's equation reduces to Laplace's equation

$$\nabla^2 \psi(r) = 0, \tag{9.10}$$

which does not necessarily mean that the electric potential is zero in that region, but rather that it is at an inflexion point.

9.4.2 Boltzmann distributions of charge around nanoparticles

Consider a charged particle of interest (such as a nanoparticle) imbedded in a solution of ions. The distribution of charges, $\rho(r)$, at some distance, r, from the particle is now given by the sum of the concentration, c_i, of ions weighted by their unitary charges, z_i

$$\rho(r) = \sum_i z_i e c_i(r). \tag{9.11}$$

where e is the charge of an electron.

In general, the chemical potential from ionic species i at any point in space, $\mu_i(r)$, is the sum of the reference chemical potential, μ^0, the chemical potential arising from the electric potential, $z_i e \psi(r)$, and the chemical potential arising from the concentration, $kT \ln c_i(r)$.

$$\mu_i(r) = \mu^0 + z_i e \psi(r) + kT \ln c_i(r) \tag{9.12}$$

If the solution is at equilibrium (i.e., there is no net flow of ions towards or away from the particle of interest) then the chemical potential of each ionic species must be the same throughout the solution, so $\mu_i(r) = \mu_i(\infty)$ or

---

[3]The terminology around the permittivity can be confusing. The permittivity of a material, ϵ, is often expressed relative to the permittivity of a vacuum ($\epsilon_0 = 8.8541\ 10^{-12}\ F\ m^{-1}$) such that $\epsilon_r = \epsilon/\epsilon_0$.

[4]Both Poisson and Laplace were engaged in many areas of mathematics and physics and their names are therefore associated with many important concepts.

$$\mu^0 + z_i e \psi(r) + kT ln c_i(r) = \mu^0 + z_i e \psi(\infty) + kT ln c_i(\infty). \qquad (9.13)$$

At some large (infinite) distance away from that particle the concentration of ions is that of the bulk concentration, the charge distribution is net zero, and for convenience we define the potential arising from the charged particle in that region to be zero as well: $\psi(\infty) = 0$. Rearranging Equation 9.13 yields the concentration of species i in terms of the bulk concentration and the electric potential as

$$c_i(r) = c_i(\infty) e^{\left(-\frac{z_i e \psi(r)}{kT}\right)}. \qquad (9.14)$$

This is called the **Boltzmann factor** which can be substituted into Equation 9.11 to give the total charge distribution

$$\rho(r) = \sum_i z_i e N_A c_i(\infty) e^{\left(-\frac{z_i e \psi(r)}{kT}\right)}; \qquad (9.15)$$

where we have introduced Avogadro's number N_A to account for the concentration units of $mol\ m^{-3}$.

Finally, combining this with the Poisson's equation (Equation 9.9), we obtain the **Poisson–Boltzmann equation** for the potential in solution arising from the charged particle of interest as a function of the distance from the particle.

$$\nabla^2 \psi(r) = -\frac{1}{\epsilon_r \epsilon_0} \sum_i z_i e N_A c_i(\infty) e^{\left(-\frac{z_i e \psi(r)}{kT}\right)} \qquad (9.16)$$

This is a second order differential equation that can be solved for a number of limiting situations. It is, however, not easy to solve in general. We will consider three limiting cases: a solution where the ions are singly charged $(z_+ = z_- = 1)$; a particle with a small surface potential; and a very large particle which can be modeled as an infinitely large wall.

9.4.2.1 1:1 Electrolyte solutions

A solution in which the ions come from a 1:1 electrolyte (such as NaCl, KCl, NaI, etc.) requires that the bulk concentrations of negatively and positively charged ions are equal, so the Poisson–Boltzmann Equation reduces to

$$\nabla^2 \psi(r) = -\frac{e N_A c(\infty)}{\epsilon_r \epsilon_0} \left[e^{\left(-\frac{e \psi(r)}{kT}\right)} - e^{\left(\frac{e \psi(r)}{kT}\right)} \right]$$

$$= -\frac{e N_A c(\infty)}{\epsilon_r \epsilon_0} sinh\left(\frac{e \psi(r)}{kT}\right). \qquad (9.17)$$

The solution to this differential equation is

$$\psi(r) = \frac{2kT}{e} \left[\frac{e^{\left(\frac{e\psi(0)}{2kT}\right)} + 1 + \left(e^{\left(\frac{e\psi(0)}{2kT}\right)} - 1\right)e^{-\kappa r}}{e^{\left(\frac{e\psi(0)}{2kT}\right)} + 1 - \left(e^{\left(\frac{e\psi(0)}{2kT}\right)} - 1\right)e^{-\kappa r}} \right], \tag{9.18}$$

where $\psi(0)$ is the potential at the surface of the particle of interest, or the surface potential, ψ_0; and κ is given by

$$\kappa^2 = \left(\frac{e^2 N_A \sum_i z_i^2 M_i}{\epsilon_r \epsilon_0 kT}\right) = \left(\frac{2e^2 N_A c(\infty)}{\epsilon_r \epsilon_0 kT}\right), \tag{9.19}$$

where we note that the ionic strength $I = \sum_i z_i^2 M_i$ becomes 2 times the bulk concentration for a 1:1 electrolyte.

The inverse of the decay constant κ is known as the **Debye length**. It is important to note that the potential in solution depends ultimately only on the surface potential, the bulk concentration of ions in solution, and the permittivity (or dielectric constant) of the medium.

Figure 9.5 compare the variation of the potential in Equation 9.18 (actually $\psi(r)/(ze/2kT) - 1$) with the distance from the particle of interest (in units of the Debye length) at four surface potentials (10 mV, 40 mV, 70 mV, and 100 mV) with a pure exponential decay function. It is evident that the decay is nearly exponential except for short distances and at high surface potentials. This is particularly evident in the logarithmic plot in the inset where we expect an exponential decay to be linear.

9.4.2.2 Debye–Huckel approximation

The result for 1:1 electrolytes suggest that when the surface potential is small (on the order of kT/e), then the surface potential decays nearly exponentially. This can be confirmed by expanding the exponential term in the charge distribution given by Equation 9.15 as a Taylor series[5] keeping the first two terms only:

$$\rho(r) = \sum_i z_i e c_i(\infty) e^{\left(-\frac{z_i e \psi(r)}{kT}\right)}$$

$$\approx \sum_i z_i e c_i(\infty) + \sum_i z_i e c_i(\infty) \left(-\frac{z_i e \psi(r)}{kT}\right) \tag{9.20}$$

$$= \frac{e^2 \sum_i z_i^2 c_i(\infty)}{kT} \psi(r)$$

$$= -\epsilon_r \epsilon_0 \kappa^2 \psi(r).$$

---

[5]The Taylor series expansion is a well-known expansion that allows for approximation of complex functions in terms of its derivatives. In this case $exp(x) = 1 + x + x^2/2! + x^3/3! + ...$

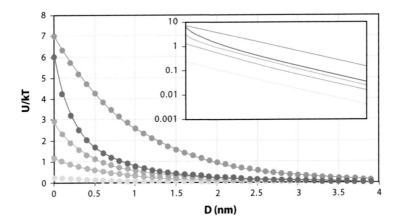

Figure 9.5: The decay of the potential as a function of distance from the surface for four surface potentials (10, 40, 70, and 100 mV light blue to red) compared to a simple exponential decay (dark blue). Inset shows the same functions on a logarithmic scale where exponential decays become linear.

The first sum in the Taylor series disappears because of electrical neutrality, and the second sum becomes the factor that ultimately defines κ.

As a consequence, the Poisson–Boltzmann equation reduces to

$$\nabla^2 \psi(r) = \kappa^2 \psi(r), \qquad (9.21)$$

which leads to the simple solution that the potential in solution arising from the particle of interest decays exponentially from the surface for any electrolyte solution as long as the surface potential is relatively small.

$$\psi(r) = \psi_0 e^{-\kappa r} \qquad (9.22)$$

This is in agreement with the result shown in Figure 9.5 for the 1:1 electrolyte seen above at low surface potentials.

For most cases, it is therefore reasonable to approximate the electric potential as decaying exponentially (the Debye–Huckel approximation) with a decay constant that in general depends only on the ionic strength of solution in which the particle is suspended. As we shall see shortly, we therefore also expect that the total repulsive interaction energy will also decay exponentially away from the surface.

9.4.2.3   Gouy–Chapman theory

In the limit where the particle is large and can be approximated by an infinitely large, flat surface, the Poisson–Boltzmann equation reduces to a one-dimensional problem. Using Laplace's Equation (Equation 9.10) we expect that the gradient of the electric potential is zero at infinite distance. Also, at the surface, we expect the gradient to depend on the surface charge, ρ_0. Thus

$$\left(\frac{\partial \psi}{\partial x}\right)_\infty = 0$$

$$\left(\frac{\partial \psi}{\partial x}\right)_0 = -\frac{\rho_0}{\epsilon_r \epsilon_0}. \tag{9.23}$$

This leads to the simpler first order differential equation

$$\left(\frac{\partial \psi}{\partial x}\right)^2 = \frac{2kT}{\epsilon_r \epsilon_0} \sum_i c_i(\infty) \left[e^{-\frac{z_i e \psi(x)}{kT}} - 1 \right]. \tag{9.24}$$

At the surface, we get

$$\left(-\frac{\rho_0}{\epsilon_r \epsilon_0}\right)^2 = \frac{2kT}{\epsilon_r \epsilon_0} \sum_i c_i(\infty) \left[e^{-\frac{z_i e \psi(0)}{kT}} - 1 \right] = \frac{2kT}{\epsilon_r \epsilon_0} \sum_i \left[c_i^{surface} - c_i^{bulk} \right]. \tag{9.25}$$

This leads to the conclusion that the surface charge, ρ_0 causes the concentration of ions at the surface to differ from that in the bulk. This concept was developed further by Stern who argued that there is a layer of counter ions that are closely associated with the surface charges and the gradient in concentrations of counter- and co-ions away from the surface leads to the gradient in electric potential.

9.4.3   The zeta-potential and its significance

The concept that there are ions closely associated with the surface of the charged particle of interest, suggest that these may move with the particle as it diffuses through the solution. These same ions would change the "effective" surface charge, decreasing it if they are counter ions and increasing it if they are ions of the same charge that for other reasons associate strongly with the surface.

If the particle is placed in a permanent electric field, the charged particle will move towards the oppositely charged pole and will reach a constant velocity, the drift velocity, u_E, which can be measured by a number of tools, such as dynamic light scattering as we will discuss further in a later section.

The relationship between the drift velocity and the "effective" surface potential, the zeta-potential, ζ, depends on how thick the film of attached ions and solvent is relative to the size of the particle. In the limit of thin films where the Debye length is much less than the radius of the particle, a, i.e., $\kappa a \gg 1$, Smoluchowski showed that

$$u_E = \frac{\epsilon_r \epsilon_0 \zeta}{\eta}, \tag{9.26}$$

where η is the viscosity of the solution.

In the other limit of very small particles or large Debye lengths, $(\kappa a < 1)$, the drift mobility is

$$u_E = \frac{2\epsilon_r \epsilon_0 \zeta}{3\eta}, \tag{9.27}$$

which can be generalized as

$$u_E = \frac{\epsilon_r \epsilon_0 \zeta}{\eta} f(\kappa a), \tag{9.28}$$

where $f(\kappa a)$ is the Henry's function, which has been approximated by Ohshima (1994) as

$$f(\kappa a) = \frac{2}{3}\left[1 + \frac{1}{2\left(1 + (2.5/\left[\kappa a\left\{1 + 2e^{-\kappa a}\right\}\right])\right)^3}\right] \tag{9.29}$$

and ranges from $2/3$ for $\kappa a \to 0$ to 1 for $\kappa a \to \infty$.

Thus measurement of the drift velocity can provide an estimate of the zeta-potential, which would correspond to the surface potential of the particle at the outer edge of the Stern layer, corresponding to the region where the adhered ions and solvent molecules are effectively associated with the particle.

The region of space where the potential is greater than zero by a significant amount is called the **Diffuse Double Layer** and its thickness is characterized by the Debye length (κ^{-1}).

The diagram in Figure 9.6 illustrates the relationship between the true surface potential, ψ_0, the zeta-potential, the diffuse double layer for a particle of radius a, and the potential in solution, $\psi(r)$.

9.4.4 The repulsive interaction potential

The electrical field felt in solution by a charged particle is seen by the above considerations to decrease exponentially with the distance from the particle with a decay constant given by the Debye length, κ. We would therefore expect the repulsive interaction potential, $U_{repulsive}$ between two charged particles to depend on their separation, D, in a similar manner, so that

$$U_{repulsive} = C_{repulsive} e^{-\kappa D}, \tag{9.30}$$

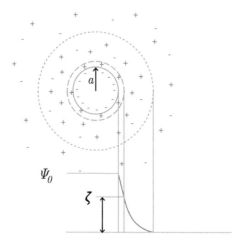

Figure 9.6: Illustration of the potentials surrounding a particle in an ionic solution

where the constant, $C_{repulsive}$ will depend on the geometry[6] as indicated in Table 9.3.

The factor, Z, in Table 9.3 can be shown to be a function of the surface potential and the properties of the intervening solvent

$$Z = 64\pi\epsilon_r\epsilon_0 \left(\frac{kT}{e}\right) tanh^2 \left(\frac{ze\psi_0}{4kT}\right). \tag{9.31}$$

Figure 9.7 illustrates the effect of increasing ionic strength on the repulsive interaction potential for two spheres of radius 100 nm and surface potential of $-40\ mV$.

9.5 FINAL STEPS TOWARDS THE DLVO THEORY

The combination of the attractive van der Waal's interparticle interactions and the repulsive ionic double layer interparticle interactions through Equations 9.4 and 9.30 lead to the Dejarguin–Landau–Verwey–Overbeck Theory also known as the DLVO Theory as shown in Equation 9.32.

$$U = U_{attractive} + U_{repulsive}$$
$$= f(geometry) \left(-A_H(\epsilon;n)/D^m + Z(\epsilon;\psi_0;I)e^{-\kappa D}\right) \tag{9.32}$$

---

[6]Note that the geometric factors are the same for the repulsive and the attractive interaction potentials as seen by comparing Tables 9.1 and 9.3.

Table 9.3: Repulsive Forces and Potentials between Surfaces of Different Geometry

Geometry	Force	$U_{repulsive}$
Two flat surfaces	$\frac{\kappa^2}{2\pi}Ze^{-\kappa D}$	$\frac{\kappa}{2\pi}Ze^{-\kappa D}$
A sphere of radius r near a flat surface	$\kappa r Z e^{-\kappa D}$	$r Z e^{-\kappa D}$
A rod of radius r near a flat surface	$\kappa^{3/2}\sqrt{\frac{r}{2\pi}}Ze^{-\kappa D}$	$\kappa^{1/2}\sqrt{\frac{r}{2\pi}}Ze^{-\kappa D}$
Two spheres of radii r_1 and r_2	$\kappa\frac{r_1 r_2}{r_1+r_2}Ze^{-\kappa D}$	$\frac{r_1 r_2}{r_1+r_2}Ze^{-\kappa D}$
Two parallel rods of radii r_1 and r_2	$\frac{\kappa^{3/2}}{\sqrt{2\pi}}\sqrt{\frac{r_1 r_2}{r_1+r_2}}Ze^{-\kappa D}$	$\frac{\kappa^{1/2}}{\sqrt{2\pi}}\sqrt{\frac{r_1 r_2}{r_1+r_2}}Ze^{-\kappa D}$
Two perpendicular rods of radii r_1 and r_2	$\kappa\sqrt{r_1 r_2}Ze^{-\kappa D}$	$\sqrt{r_1 r_2}Ze^{-\kappa D}$

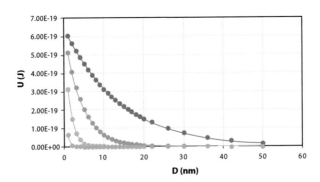

Figure 9.7: Variation of the repulsive interaction potential for two spheres at different ionic strengths

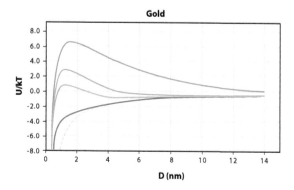

Figure 9.8: Variation DLVO interaction potential for two gold spheres at increasing ionic strengths from blue to red.

The overall interparticle interaction potential is therefore a balance between an exponentially decaying function which is particularly sensitive to the ionic strength, I, of the solution and the surface potential, ψ_0 of the particles and a hyperbolic function which is relatively insensitive to the solution properties. At low ionic strength, the particle surface potential is not shielded effectively by the ions in solution and hence the repulsive interactions extend to long distances and dominate the total interparticle interaction potential. Under these circumstances, the particles tend to stay dispersed and the suspension is relatively stable. At high ionic strength, the particle surface potential is shielded both by the ions in the Stern layer and by the ions in the diffuse double layer even at short distances and hence the attractive interparticle interaction potential dominates. Under these circumstances, the particles will tend to either flocculate or aggregate and sediment out of solution. The suspension is unstable. Correspondingly, a large surface potential (either positive or negative) will stabilize a suspension of nanoparticles, while a small surface potential will destabilize the suspension.

For two identical, spherical particles of radius, r, Equation 9.32 reduces to

$$ U = \frac{r}{2}\left(-\frac{A_H}{6}\frac{1}{D} + Ze^{-kD}\right), \tag{9.33} $$

where the Hamaker Constant, A_H is given by Equation 9.7, Z is given by Equation 9.31, and κ is given by Equation 9.19.

Figures 9.8 and 9.9 illustrate, respectively, the effects of ionic strength on the interaction potentials (in units of kT) between two $10-nm$ gold nanoparticles (surface potential of $-40\ mV$) and two $100-nm$ membrane vesicles (surface potential $-10\ mV$) as a function of their separation in nm.

At low ionic strengths, the repulsive interaction potential dominates and the particles would not approach each other closely and the energy required

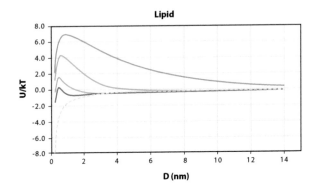

Figure 9.9: Variation of the repulsive interaction potential for membrane spheres at increasing ionic strengths from blue to red.

to surpass the energy barrier to lead to aggregation (at distances closer than 1 nm) is several times kT. At intermediate ionic strengths, the attractive interaction potential is large enough to create a low-energy minimum at 5-8 nm for the case of the gold nanoparticles and 3 nm for the membrane vesicles. In both cases these minima are at energies on the order of kT and hence these represent reversible interactions, corresponding to flocculation. These are not stable for the gold nanoparticles, but can become important for the membrane vesicles, which are known to create flocculates that can readily be reversed. At sufficiently high ionic strengths, the barrier to strong associations at distances below 1 nm is less than kT and hence irreversible aggregations can readily form.

9.6 LIMITATIONS AND APPLICATIONS

While the general principles of the DLVO theory phenomenologically explains the behavior of nanoparticles in solution, it is an approximation with a number of limitations. Among these is the general assumption that the distances of separation are small relative to the size of the objects. In classical colloid theory, where particles were generally greater than about 100 nm, this is a good assumption, but for smaller nanoparticles, this is a problem. A more general expression of the interaction of two spherical nanoparticles (radius r_1 and r_2) with surface potentials ψ_1 and ψ_2 in water as a function of their separation, D, has been developed (see for example Zhang, Chapter 2)

$$U_{attractive} = -\frac{A_H}{6}\left(\frac{8r_r^2}{D(8r_r+D)} + \frac{8r_r^2}{(4r_r+D)} + ln\frac{D(8r_r+h)}{(4r_r+h)^2}\right)$$

$$U_{repulsive} = 4\pi\epsilon\epsilon_0 r_r\left(\psi_1\psi_2 e^{-\kappa D} - \frac{1}{4}\left(\psi_1^2 + \psi_2^2\right)e^{-2\kappa D}\right), \qquad (9.34)$$

where r_r is the reduced particle radius $r_r = r_1 r_2/(r_1 + r_2)$, and the other parameters are as defined previously.

It is possible to show (see problem assignment) that these equations reduce to the DLVO theory as expressed by Equation 9.33 when $r_r \gg D$.

Beyond these geometrical limitations, the DLVO Theory is limited to consideration of the attractive van der Waal's interactions and the repulsive electrostatic interactions. A number of refinements can be introduced by considering "non-DLVO" interactions such as the effect of surface roughness on the particles, which becomes increasingly important as the particle size decreases. Consider, for example that most metallic nanoparticles are really small crystals, with surfaces determined by the crystallographic surfaces, so that there are flat regions and sharp edges. Such refinements lead to what is frequently called "Extended-DLVO" or EDLVO or XDLVO theories.

One of the common uses of the DLVO (or EDLVO) theory is to explain or predict the kinetics of aggregation of nanoparticles into larger colloid structures. In general, there are two limits of the kinetics to be considered in regard to that application of the DLVO theory — the limit where the particles stick with high efficiency when they encounter each other and the limit where the particles need to collide many times before they stick together. The first of these limits correspond to a "diffusion" limited process, in which the energy barrier to association in the primary potential well is low and as a consequence every encounter lead to an aggregation of the particles. The second of these limits correspond to a "reaction" limited process, in which the energy barrier to the association in the primary potential well is high and as a consequence the probability of aggregation of the two particles in any given encounter is low. In both cases, the energy barrier to aggregation, as given by the DLVO theory (or its extensions), can be viewed as an activation energy for the aggregation kinetic process as illustrated in Figure 9.10.

If the activation energy for aggregation is low, the process of aggregation is diffusion limited and the outcome of the aggregation process is a loosely formed aggregate (also called a dendritic aggregate) as illustrated schematically in Figure 9.11A. On the other hand, if the activation energy for aggregation is high, the process of aggregation is reaction limited and the outcome of the aggregation is a more compact aggregate as illustrated schematically in Figure 9.11B.

We saw that the DLVO theory predicts that as the particle size decreases, the energy barrier is expected to decrease proportionally and hence we would

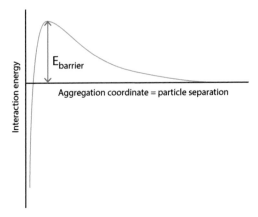

Figure 9.10: Illustration of the energy barrier as an activation energy for aggregation.

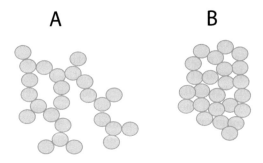

Figure 9.11: Illustration of a dendritic aggregate (A) and a compact aggregate (B).

expect the activation energy to decrease and the aggregation rate to increase correspondingly. Likewise, we might expect the smaller particles to form more dendritic-like aggregates. However, the smaller the particle, the more the surface contours become important and the DLVO theory becomes less accurate. These predictions are therefore not always observed.

The DLVO theory also predicts that the energy barrier decreases as the ionic strength, or the concentration of ions in the solution increases. One may therefore expect that there will be a range of concentrations of salts where the aggregation (or coagulation) process is greatly enhanced. The concentration at which the aggregation is truly diffusion limited is the concentration at which

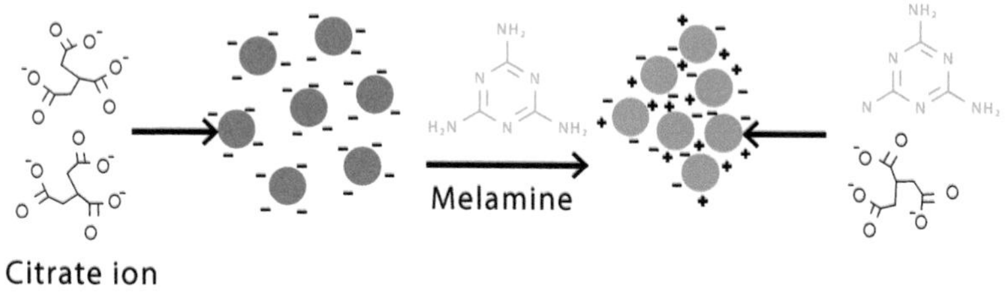

Citrate ion

Melamine

Figure 9.12: The use of melamine as indicator of aggregation. The shift in color from red to blue will occur as the gold nanoparticles aggregate [adapted from Li et al.].

the interaction potential disappears and this is referred to as the "critical coagulation concentration" or the CCC. This will occur at a distance, D_{CCC}, where

$$U(D_{CCC}) = 0 \quad \left(\frac{dU}{dD}\right)_{D_{CCC}} = 0. \tag{9.35}$$

A simple example of the measurement of the CCC for gold nanoparticles was used by Li, Yu, and Zou in 2014 to determine the concentration of melamine in solution. They used the adhesion of melamine (positively charged) to the gold nanoparticle (negatively charged) to cause coagulation of the gold nanoparticles, which in solution are red, to form aggregates that are blue (as a consequence of a shift in the plasmon resonance). The change in color is the indicator of aggregation, as indicated in Figure 9.12. The assay then includes monitoring how much of a salt needs to be titrated into a mixture to cause the color change at various melamine concentration to create a standard curve, as in Figure 9.13. The concentration of salt added to change the color from red to blue would be an indicator of the concentration of melamine in solution.

The application of DLVO theory in its simplest form is also limited by the shape of the nanoparticle. Specifically, we expect that spheres, rods, and flakes of nanoparticles will have different aggregation behavior. The distance dependence of the interaction potential will depend on the orientation of these nanoparticles with respect to each other (as is evident from Tables 9.1 and 9.3) and the dependence on their separation is more severe as the surfaces become

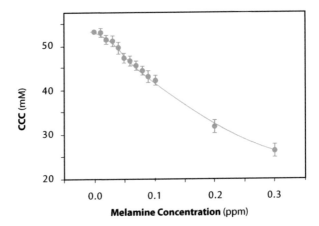

Figure 9.13: The concentration of salt that leads to color change (the CCC) as a function of melamine concentration [adapted from Li et al.].

flatter. We therefore may expect that two flat surfaces will attract each other more effectively.

The concepts of molecular self-assembly uses the differential adhesion interaction at different surfaces to create supramolecular structures of defined, and predictable geometries.

IV

Understanding fluctuations in small systems

Introduction to Part IV

Nanoscale systems, by their very nature, are small and hence there is a relatively small number of molecules contained within them along critical dimensions. We saw in Table 7.2 that spherical particles of a few nanometers contain on the order of a few hundred to a few thousand atoms and many, if not most, of these are on the surface. Similarly, the number of molecules within a thin film will define the thickness of the film which in turn determines the properties of the film. Or the cross-section dimensions of a nanowire is related to the number of molecules that can fit within it.

Since most of the properties of such nanoscale materials depend on their dimensions, we can expect that there will be a range of properties if there is a range of dimensions in any given system. Experimentally, it may be difficult to control assembly of nanoparticles to provide one and only one thickness of a film or a wire or radius of a sphere. We therefore expect that there will be a range of dimensions characterized by some distribution. We will therefore have to worry not only about the average properties of a system of nanoparticles, but also about the range of these properties.

As an example, consider the energy associated with a transition of an electron from the highest occupied electronic state to the lowest unoccupied electronic state within a nanoparticle such as a quantum dot. For a single particle of a defined size, the energy is well defined and there is a single wavelength of light associated with the absorption or emission. For a collection of particles of different sizes, there will be a range of energies and a corresponding range of wavelengths in the absorption or emission spectra.

We often attempt to study or use nanoscale systems in small volumes or at low concentrations which in turn means that the number of nanoscale systems being observed at any given time is small. For example, we may trap a small number of catalytic nanoparticles within a matrix of larger particles to stabilize them and to incorporate them in a functional unit. In this case, the properties of the larger matrix particles will depend on the number of nanoparticles within them. Since the number of nanoparticles is likely to vary among the matrix particles, there will also be a range of properties among these matrix particles.

The point of these examples, is that within the world of nanoscale materials, we need to understand the impact of having distributions, variations, or fluctuations in properties of the materials.

The purpose of this part is to understand the origin of the distributions, how they can be characterized, and how they can be utilized. In this section, the first chapter (Chapter 10), will review some of the basic properties of distributions in general and discuss some particularly important distributions. The second chapter (Chapter 11) will provide the fundamentals of Statistical Thermodynamics needed to introduce the concept of partition functions in isolated, closed, and open systems as defined in Section 6.1. The third chapter (Chapter 12) introduces the concept of fluctuations in energy in closed systems and fluctuations in energy and numbers of molecules in open systems. Finally, the fourth chapter (Chapter 13) demonstrates how fluctuations in open systems can be used to determine concentrations of nanoscale system as well as the kinetics of the processes that give rise to the fluctuations that are being observed.

CHAPTER 10

Basic statistical concepts

The purpose of this chapter is to review some basic statistical concepts that are encountered repeatedly in the following chapters.

Webster's Encyclopedic Dictionary of the English Language (Canadian Edition, 1988) defines the field of statistics as:

> **Statistics n. the collection and study of numerical data, esp. as the branch of mathematics in which deductions are made on the assumption that the relationship between a sufficient sample of numerical data are characteristic of those between all such data.**

As this definition implies, statistics is about analyzing samples of data and making deductions that should be representative of the real system from which we sample. When we sample, we normally encounter that the data distribute in a manner that is characteristic of the system we study. From analysis of the distribution we may obtain information about the system. For our purposes, we focus on understanding various distributions and what information these distributions contain.

10.1 DISTRIBUTIONS AND THEIR CHARACTERISTICS

10.1.1 Measured distributions and distribution functions

Consider a number of individual measurements of a property, such as the diameter, d, of a each of the nanoparticles in an electron microscopy image as in Figure 10.1. It is evident from the image that there would be a range of diameter measurements and there would be a few small diameters, a few large diameters, and a number of diameter measurements in between. In this example, there are $N = 103$ distinguishable[1] nanoparticles that can be measured with a ruler in 1 mm increments (on an enlarged image where 1 mm = 3 nm).

---

[1]There are a number of particles that have aggregated and their individual sizes cannot be reliably measured.

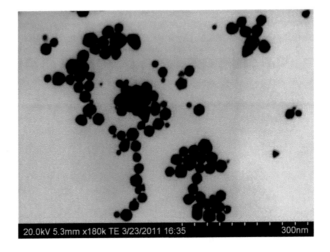

Figure 10.1: Electron microscopy image of gold nanoparticles measured in the authors laboratory.

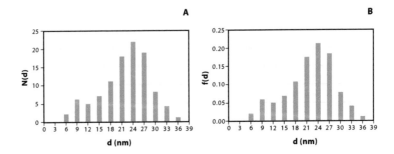

Figure 10.2: Absolute (A) and frequency (B) distribution of diameters from Figure 10.1.

The number of particles, $N(d_i)$, of a particular diameter, d_i, is shown in Figure 10.2A for each diameter; this shows the **absolute distribution** of diameters. When we divide each of the counts by the total number of particles measured, we **normalize** the distribution and calculate the frequency, $f(d_i) = N(d_i)/N$ with which a particular diameter is measured. This **frequency distribution** is shown in Figure 10.2B and looks identical to the original distribution with the difference being that the vertical axis is in different units.

The frequency distribution is useful since the area under the entire curve, through the normalization, is unity. This is illustrated by creating the **cumu-**

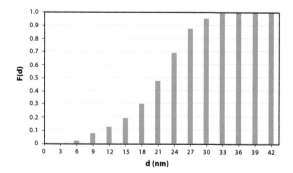

Figure 10.3: Cumulative distribution function corresponding to the frequency distribution in Figure 10.2B.

lative distribution function[2] as shown in Figure 10.3, where the value at a particular diameter, d, corresponds to the sum of all the frequencies for sizes smaller or equal to that diameter, i.e.,

$$F(d) = \sum_{d_i=0}^{d_i=d} f(d_i) = \sum_{d_i=0}^{d_i=d} \frac{N(d_i)}{N} = \frac{\sum_{d_i=0}^{d_i=d} N(d_i)}{\sum_{d_i=0}^{d_i=d_{max}} N(d_i)} \qquad (10.1)$$

This distribution function can also be interpreted as the area under the distribution curve from zero diameter up to the diameter of interest. For example, in this distribution it appears that close to half of the nanoparticles have a diameter less than or equal to 21 nm.

The frequency with which a particular diameter is measured can be used to estimate the probability that a randomly selected nanoparticle will have that diameter. The measured frequency distribution is therefore an approximation of an underlying **probability distribution** which can be used to predict the likelihood of a particular outcome. For example, the frequency distribution in Figure 10.2B would allow us to estimate that the probability of finding a $30 - nm$ diameter nanoparticle will be about 0.07 (or about 7%).

Likewise, the measured cumulative distribution function is an approximation of the corresponding **probability distribution function**, which is useful for predicting the probability of finding particles smaller than (or greater than) a particular diameter.

The measured distribution and distribution functions are always discrete distributions or distribution functions, meaning that the measurements are grouped within specific ranges of the measurements. In this example, the diameters are set at multiples of 3 nm so that the frequency of a diameter

---

[2]Note the distinction between a distribution and a distribution function. The distribution shows the individual frequencies while the distribution function shows the sum of the frequencies.

of 12 nm actually represents all measurements of diameters between 10.5 and
13.5 nm. This discreteness arises because there is a limit on the **precision**[3]
with which the diameter can be measured. Accordingly, the corresponding
probability distribution and probability distribution function are also discrete.

In contrast to measured distributions and distribution functions, theoretical distributions can be either discrete or continuous. We will discuss this
further below.

10.1.2 Characteristics of a distribution

A distribution can in principle be fully characterized by its moments. The
general definition of the n-th moment, μ_n, for a discrete distribution of a
parameter, such as the diameter, d_i, is given by

$$\mu_n = \sum_i (d_i - d')^n f(d_i), \tag{10.2}$$

where d' is the value of the parameter about which the moment is calculated.
If $d' = 0$, the moment is calculated about the zero value of the parameter.

In all cases, the zero'th moment of a normalized distribution is 1 since it
is simply the sum of the frequency distribution.

In the case where $d' = 0$, we obtain what are sometimes termed the **raw
moments**. The first raw moment of the normalized distribution is called the
mean, μ.

If the moments are calculated about the mean, then they are called the
central moments. The first central moment is then equal to zero. The second
central moment is the **variance**, the third central moment is the **skewness**,
and the fourth central moment is the **kurtosis**:

---

[3]Remember that precision refers to how reproducibly a sample can be measured. How
good are we at measuring?

$$\mu = \sum_i (d_i)^1 f(d_i)$$

$$\mu_1 = \sum_i (d_i - \mu)^1 f(d_i) = 0$$

$$\mu_2 = \sum_i (d_i - \mu)^2 f(d_i) = \sigma^2$$

$$\mu_3 = \sum_i (d_i - \mu)^3 f(d_i)$$

$$\mu_4 = \sum_i (d_i - \mu)^4 f(d_i)$$

$$\dots$$

$$\mu_n = \sum_i (d_i - \mu)^n f(d_i). \tag{10.3}$$

The mean is also known as the average of the distribution and represents the expected value or the expectation value of the parameter, d. It therefore represents the value that we expect to observe if we make a random measurement. Unless the distribution is symmetric, the mean is not necessarily the same as the **mode** of the distribution, which is the most frequently observed value (i.e., the peak of the distribution). Likewise, for asymmetric distributions the mean is different from the **median**, which represents the value of the parameter where half of the measurements are smaller and half are larger, i.e., the point at which the distribution function is 0.5. In the case of the distribution in Figure 10.2B, the mean is 21.9 nm, the mode is 24 nm, and the median is 24 nm.

The variance is a measure of the width of the distribution. If the distribution is narrow, the variance is small, but if the distribution is wide, the variance is large. The square root of the variance is called the **standard deviation** and represents a measure of how much the distribution differs from the mean. The standard deviation is distinct from the **mean deviation**, which is the same as the first central moment, μ_1, which we see from Equation 10.3 is always zero.

The higher order moments describe the asymmetry of the distribution. A negative skewness indicates that the distribution is skewed towards lower values of the parameter of interest while a positive skewness indicates that the distribution is skewed towards higher values of the parameter of interest. In the distribution in Figure 10.2, the skewness is negative because of the relatively high frequency of diameter measurements between 3 and 12 nm.

Theoretical probability distributions, $p(x)$, and probability distribution functions, $P(x)$, are frequently continuous, which simply means that there is a probability associated will all values of the parameter of interest. The more precisely we can measure a distribution, the more it will approach the continuous distribution. The advantage of the continuous distribution framework is that we can replace the sums in our calculations by integrals. Thus the definition of the moments about a particular value as in Equation 10.2 become

$$\mu_n = \int_{\infty}^{\infty} (x - x')p(x)dx. \tag{10.4}$$

The link between the continuous probability distribution, $p(x)$, and the raw moments, μ'_n, is provided by the moment generating function, $M(t)$, as indicated in Equation 10.5

$$M(t) = \int_{\infty}^{\infty} e^{tx} p(x)dx = 1 + \mu'_1 t + \frac{1}{2!}\mu'_2 t^{+} \frac{1}{3!}\mu'_3 t^3 + \dots, \tag{10.5}$$

which means that if we know all the raw moments, we can recover the probability distribution through an inverse transform of Equation 10.5.

10.1.3 Some important distributions

There are a number of well-known distributions that appear repeatedly because they represent realistic scenarios. By comparing experimental data to these known distributions we can get information about the underlying physical processes that generated the distribution in the first place. We will examine a few of these: the binomial, the Poisson, the normal (Gaussian), the log-normal, and the Boltzman distributions.

10.1.3.1 The binomial distribution

The binomial distribution arises from repeated and independent experiments in which the outcome is either of two outcomes (hence the binomial label). This is a Bernoulli trial in which a random experiment results in either success or failure. If the probability of success in any single experiment is p, then the probability of failure is $q = 1 - p$. The binomial distribution describes the probability, $b(k; n, p)$ of obtaining exactly k successful outcomes in n experiments, and can be shown to be

$$b(k; n, p) = \binom{n}{k} p^k q^{n-k}, \tag{10.6}$$

where $\binom{n}{k} = \frac{n!}{k!(n-k)!}$.

Recall that the symbol ! represents the factorial, such that $n! = n \cdot (n-1) \cdot (n-2) \cdot (n-3) \dots 3 \cdot 2 \cdot 1$. For example $5! = 5 \cdot 4 \cdot 3 \cdot 2 \cdot 1$. By definition, $0! = 1$.

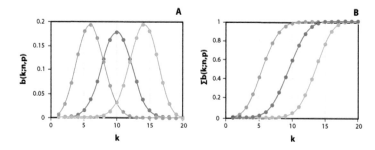

Figure 10.4: Binomial distributions (A) and distributions functions (B) for $n = 20$ and $p = 0.3$ (blue), 0.5 (red), and 0.7 (orange).

The classical example of a Bernoulli experiment that leads to a binomial distribution is that of repeated coin tosses, where success could be heads and failure could be tails. In this case, the probability of success equals the probability of failure and $p = q = 0.5$.

The mean of the binomial distribution is $\mu = np$; the second central moment, and the variance, is $\mu_2 = \sigma^2 = np$; and the third central moment is $\mu_3 = npq(1 - 2p)$ which is zero when $p = 0.5$, negative when $p > 0.5$, and positive when $p < 0.5$. The binomial distribution is therefore generally skewed, except when the probability of success and failure are the same. This is illustrated in Figure 10.4 for values of $p = 0.3$, $p = 0.5$, and $p = 0.7$ and $n = 20$.

As we will see later, we will be interested in the ratio of the variance to the square of the mean, and we see that for the binomial distribution, this quantity is $\frac{\sigma^2}{\mu^2} = \frac{q}{p}\frac{1}{n} = \frac{q}{\mu}$, i.e., proportional to the inverse of the number of measurements as well as the inverse of the mean np and is equal to $1/n$ when $p = q$.

Note that the binomial distribution is a discrete distribution since it is concerned with the number of times an experiment succeeds or fails.

10.1.3.2 The Poisson distribution

The Poisson distribution arises from a **Poisson process**, which we also generally consider a **random process**. Formally, the Poisson process originates from a **Markov process**, which is a process where the transition from one state to the next state is independent of the past and depends only on the present state and the probability a transition will take place. The Poisson process must obey several conditions (Weisstein, Eric W. "Continuous Distribution." From MathWorld — A Wolfram Web Resource. http://mathworld.wolfram.com/ContinuousDistribution.html).

1. The number of events within a distinct, small period of time are independent; this is called the independent increment property.

2. The probability of a single event occurring in a small period, usually of time, is proportional to the rate of the process, often designated by λ, which is therefore equal to the expected number of events per unit time.

3. The number of events occurring in a time period between t and $t + s$ follows a Poisson distribution.

This description of the Poisson process suggests that it is useful for observing arrival events, such as arrival of photons at a detector.

The probability, $p(k; \mu)$, of observing exactly k events, given that the mean of the Poisson distribution is μ, is given by

$$p(k; \mu) = \frac{\mu^k e^{-\mu}}{k!}. \tag{10.7}$$

The Poisson distribution therefore describes the number of events that we may observe in any particular time period if we know what the mean number of events are, for example from a measurement over many time periods.

Another approach[4] to describing the Poisson distribution is to consider a Poisson experiment with the following properties:

1. The experiment results in outcomes that can be classified as successes or failures (it is a Bernoulli experiment).

2. The average number of successes, μ, that occur in a particular region is known.

3. The probability that a success occurs is proportional to the size of the region.

4. The probability that a success occurs in an extremely small region is virtually zero.

In this description, it is easier to understand that the "region" could represent a number of different physical parameters, such as the length or area of a detector, the volume of a solution, or a distinct measurement time.

The mean, μ, of the Poisson distribution is contained within the definition; the second central moment (or the variance) is equal to the mean, so $\mu_2 = \sigma^2 = \mu$; and remarkably, the third central moment is also equal to the mean, so $\mu_3 = \mu$, which is always positive, so the distribution is always skewed toward larger values as illustrated in Figure 10.5. While the second and third central moments are both equal to the mean, this trend does not continue since the higher order central moments are not equal to the mean.

---

[4] http://startrek.com/lesson2/poisson.aspx

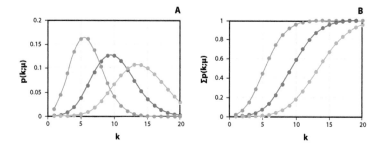

Figure 10.5: Poisson distributions (A) and distributions functions (B) for means $\mu = 6$ (blue), 10 (red), and 14 (orange).

In turn, the ratio of the variance and the square of the mean now becomes $\frac{\sigma^2}{\mu^2} = \frac{1}{\mu}$, which is also inversely proportional to the mean. We will encounter this property again later.

The Poisson distribution is also a discrete distribution.

10.1.3.3   The normal distribution

The normal distribution, also at times called the Gaussian distribution, describes the frequency with which one expects to observe the average value of a large number of independent measurements. As an example, consider tossing a coin ten times and recording the number of heads observed. Repeating this process ten times will create an experimental distribution approaching the expected binomial distribution with an average value that will be an estimate of the true average. Now, consider creating a hundred such experimental distributions. These hundred average values are expected to distribute as a normal distribution. In other words, the uncertainty (or error) in determining the average value in any given experiment is expected to follow a normal distribution. This concept is an expression of a **central limit theorem** which holds for a large number of experimental measurements. The example provided here is the foundation for using the normal distribution in general error analysis.

The normal distribution with a mean of μ and a variance of σ^2 is defined as the frequency with which we observe the value k as

$$f(k) = \frac{1}{\sigma\sqrt{2\pi}} e^{\left(\frac{\frac{1}{2}(k-\mu)^2}{\sigma^2}\right)}, \tag{10.8}$$

where the prefactor ensures that the frequency distribution is normalized.

The normal distribution is symmetric and centered around the mean value,

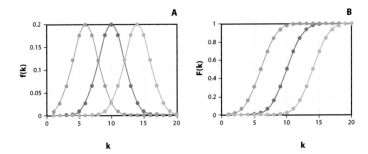

Figure 10.6: Normal distributions (A) and distributions functions (B) for means $\mu = 6$ (blue), 10 (red), and 14 (orange) and a standard deviation $\sigma = 2$.

which is also the maximum value, as seen in Figure 10.6 for three different means with the same variance.

Note that Equation 10.8 is written as a discrete distribution. In general, the normal distribution is a continuous distribution expressed as

$$p(x) = \frac{1}{\sigma\sqrt{2\pi}}e^{\left(\frac{\frac{1}{2}(x-\mu)^2}{\sigma^2}\right)} \tag{10.9}$$

The standard normal distribution arises when the mean is zero and the variance is unity, in which case Equation 10.9 reduces to $p(x) = \frac{1}{\sqrt{2\pi}}e^{\left(\frac{1}{2}x^2\right)}$. The effect of changing the variance for a fixed mean value is illustrated in Figure 10.7.

10.1.3.4 The log-normal distribution

The log-normal distribution arises when the logarithm of a variable yields a normal distribution. This arises when the process is a result of products, rather than sums, of independent measurements. The formulation is therefore the same as Equation 10.9 with the variable x replaced by its natural logarithm lnx, with the limitation that $x > 0$.

$$p(x) = \frac{1}{\sigma\sqrt{2\pi}}e^{\left(\frac{\frac{1}{2}(lnx-\mu)^2}{\sigma^2}\right)} \tag{10.10}$$

and as illustrated in Figure 10.8.

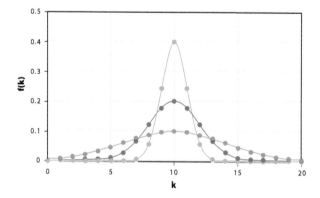

Figure 10.7: Normal distributions (A) for means $\mu = 10$ and standard deviations $\sigma = 1$ (orange), 2 (red), and 4 (blue).

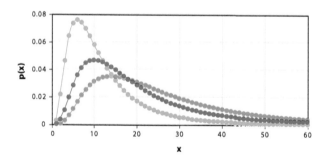

Figure 10.8: Log-normal distributions (A) and distributions functions (B) for means $\mu = 6$ (orange), 10 (red), and 14 (blue) and a standard deviation $\sigma = 2$.

10.1.3.5   The Boltzmann distribution

The Boltzmann distribution describes the fraction of molecules, $\frac{n_i}{N}$, occupying the energy level, ϵ_i, with degeneracy, g_i.

$$\frac{n_i}{N} = \frac{g_i e^{-\frac{\epsilon_i}{kT}}}{q}, \tag{10.11}$$

where

$$q = \sum_j g_j e^{-\frac{\epsilon_j}{kT}} \tag{10.12}$$

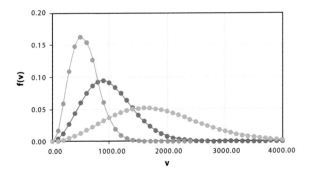

Figure 10.9: Maxwell–Boltzmann velocity distribution for neon at $T = 100\ K$ (blue), $300\ K$ (red), and $1000\ K$ (orange).

ensures that the distribution is normalized. As we shall see later, the quantity q is called the **molecular partition function**, since it is the sum of the partition of all the molecules across all the possible energy states available.

If the energy levels are quantized, the Boltzmann distribution is a discrete distribution.

The Maxwell–Boltzmann distribution of speeds is a special case of the Boltzmann distribution which describes the fraction of molecules, $f(v)$, in an ideal gas with a particular speed, v, which corresponds to a kinetic energy of $E_{kin} = \frac{1}{2}mv^2$ where m is the mass of the molecule. This is a continuous distribution which has the form

$$f(v) = \left(\frac{m}{2\pi kT}\right)^{\frac{3}{2}} 4\pi v^2 e^{-\frac{\frac{1}{2}mv^2}{kT}}. \qquad (10.13)$$

The Maxwell–Boltzmann distribution of speeds for a gas of neon molecules ($m = 3.35 \times 10^{-26}\ kg$ is shown in Figure 10.9 at three different temperatures.

10.1.4 Limits of the binomial distribution

The details of the probability distributions, such as the binomial and Poisson distributions, are quite different and they represent distinctly different physical phenomena, but they share the underlying premise that each experiment is an independent event. As a consequence, there are limits where different distributions approach each other.

If we consider a large number of Bernoulli experiments, N, with a probability of success p, then we know that the probability distribution is the binomial distribution (Equation 10.6. However, for a very large number of experiments (or samples), that is as $N \to \infty$, and if the probability of success is small, that is $p \to 0$, it is possible to show that the binomial distribution transforms into the Poisson distribution (Equation 10.7) with the constraint that the mean $\mu = N \cdot p$. This is the *law of rare events* and it is a useful concept since it implies that when the probability of successes are small, the Poisson distribution will be a good approximation.

Correspondingly, for a large number of Bernoulli experiments in which the probability of success is neither close to zero or to unity, then as $N \rightarrow \infty$ the binomial distribution approaches the normal distribution (Equation 10.8). This is a further reflection of the central limit theorem, and it is useful since it confirms the general usefulness of the normal distribution for a lot of experimental situations. The normal distribution, particularly the standard normal distribution, is also useful because it is fairly easy to manipulate and work with and it has a continuous distribution analog.

10.2 INHERENT VARIABILITY

One of the benefits of applying a known distribution to model experimental data is that it may provide additional insight into the source of the underlying uncertainty. In general, we are concerned with several sources of uncertainty.

There can be **inherent variability** in the system under study, such as the size distribution associated with the nanoparticles shown in Figure 10.1. In that case the distribution may provide information about the underlying physical events that determine the growth rate of nanoparticles. Is it nucleation and growth or Ostwald ripening or some other process?

There can be limits on the **reproducibility of sampling** in either space or time. This would lead to a variation that arises from the technique of interrogation rather than from an inherent variability of the system. This can arise for limitations in the precision with which we measure and in that case the distribution tells us about the error in measurement.

There can also be uncertainties arising from the analysis of data. There is always noise in any measurements and there may be variability in how well the data can be fit to models used for analysis. For example, the measured quantity may be x, but the parameter used to fit in models may be $1/x$. The distribution of x is different from the distribution of $1/x$ and more importantly, the mean of the distribution in x will be different from the inverse of the mean of the distribution of $1/x$, that is, $\mu(x) \neq 1/\mu(1/x)$. The question will then arise as to which is the appropriate distribution to analyze and make conclusions from. This in turn will depend on which of the parameters give rise to the greatest source of uncertainty. For example, if the greatest source of error is likely to arise from reproducibility in measurement of the parameter x, it is likely that the distribution in x is the most relevant distribution. However, if the greatest source of error is likely to arise from the inherent variability in the system, which is related to the model parameter $1/x$, then this is likely that the distribution in $1/x$ is the most relevant distribution.

These concepts will be developed further in some of the problem assignments.

CHAPTER **11**

Partition functions

The purpose of this chapter is to build on our understanding of distribution to consider the distributions of energy in various thermodynamics systems and thereby introduce the concept of partition functions, which in general terms are reflections of how the energy is partitioned across the components of a system. The partition functions are a key link between the molecular nature of materials and the thermodynamic properties of these materials. They therefore link the world of Statistical Mechanics with the world of Thermodynamics.

11.1 ISOLATED SYSTEMS

Recall from Section 6.1 that we defined an isolated system as one which cannot exchange energy or mass with its surroundings. The first and second laws of thermodynamics were defined in terms of the isolated systems and this allowed us to develop an extensive thermodynamic framework as illustrated in Chapter 6.

We can define an isolated system in terms of the three state functions: N, V, and E. In other words, an isolated systems contains a fixed number of molecules (fixed mass) in a defined volume with a fixed total energy. This does not, however, tell us anything about how the energy is distributed among the molecules in the system. Do all molecules have the same energy? Do some have higher than average energy and others lower than average energy? What is the distribution of energies?

We have seen in Section 10.1.3.5 that for an ideal gas, there is a distribution of speeds and hence a distribution of kinetic energies of the molecules as given by the Maxwell–Boltzmann distribution (Equation 10.13 and Figure 10.9). Since there are no intermolecular interactions in an ideal gas, this distribution describes the fraction of molecules with a particular total energy.

The Maxwell–Boltzmann distribution is a special case of the Boltzmann distribution (Equations 10.11 and 10.12) and we can therefore generalize and infer that for an isolated system, the probability of finding a molecule with

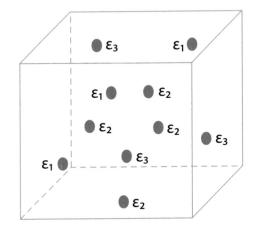

Figure 11.1: Illustration of the energy distribution in an isolated system.

the energy ϵ_i is the same as the fraction of molecules, n_i/N, with energy, ϵ_i and is given by

$$p(\epsilon_i) = \frac{n_i}{N} = \frac{g_i e^{-\frac{\epsilon_i}{kT}}}{q}, \qquad (11.1)$$

where

$$q = \sum_j g_j e^{-\frac{\epsilon_j}{kT}} \qquad (11.2)$$

with the constraints that $N = \sum_i n_i$ and $E = \sum_i \epsilon_i$.

The concept is illustrated in Figure 11.1 where there are 10 molecules with three at energy ϵ_1, four at energy ϵ_2, and three at energy ϵ_3. As a consequence we could calculate $q = 3e^{-\epsilon_1/kT} + 4e^{-\epsilon_2/kT} + 3e^{-\epsilon_3/kT}$ and $n_1/N = 3e^{-\epsilon_1/kT}/q$.

In an isolated system we are therefore concerned with the energy distribution **among** the molecules within the system. We see that this distribution depends only on the energy states available in the system and on the temperature of the system. If the temperature decreases, we might expect the distribution to be narrower since fewer high energy states are accessible. On the other hand, if the temperature increases, we might expect the distribution to broaden as more high energy state are accessible. This was illustrated in Figure 10.9 for the special case of the Maxwell–Boltzmann distribution in an ideal gas. Note that in general we expect the distribution to be quite broad.

The parameter, q, is called the **microcanonical partition function**. As we shall see later, the microcanonical partition function is related to key thermodynamic parameters for isolated systems. This is because it represents the sum of the partitioning of energy across all the molecules in the system. In

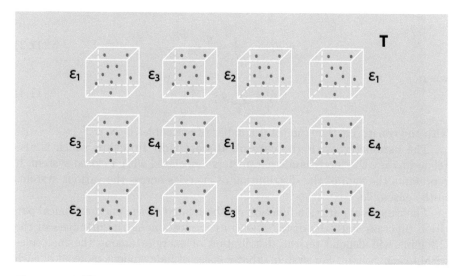

Figure 11.2: Illustration of an ensemble of closed systems in a common thermal bath at temperature T.

principle, q can be calculated if we know all of the energy levels, ϵ_i, from, for example, quantum mechanical calculations of the system. In practice, however, this is is a challenge since in most cases we are dealing with a very large number of molecules ($\approx 10^{-20}$ or more).

11.2 CLOSED SYSTEMS

Closed systems are defined as systems that allow energy, but not mass, to flow across the boundaries. They are therefore characterized by the three state functions of N, V, and T, which implies that the system is in thermal equilibrium with a large bath at constant temperature.

If we examine any example of a closed system, we will find it has a total energy E_i. However, if we examine another example of the same closed system, it may have different total energy E_j. In addition to knowing what the distribution of energy among molecules in the system may be, the distribution of total energy among different systems becomes important. In order to address that, we consider an **ensemble of systems**, which means that we examine a large number of systems of identical N, V, and T and ask how they differ with respect to the total energy. This concept of an ensemble of systems is illustrated in Figure 11.2.

Each of the members of the ensemble are representative of the closed system but they differ in their total energy. The distribution that describes the probability of finding a particular member of the ensemble with the energy E_i is measured by the **fraction of ensembles**, n_i/N that has that energy, which also follows a Boltzmann distribution

$$p(E_i) = \frac{n_i}{N} = \frac{g_i e^{-\frac{E_i}{kT}}}{Q}, \qquad (11.3)$$

where

$$Q = \sum_j g_j e^{-\frac{E_j}{kT}} \qquad (11.4)$$

with the constraint that the total number of ensembles is $N = \sum_i n_i$.

The parameter, Q, is the **canonical partition function** which is also related to key thermodynamic parameters associated with a closed system. It represents the sum of the distribution of energies across the various systems in the ensemble.

The canonical partition function can be related to the microcanonical partition function since the distribution of energies among the members of the ensemble will depend on the distribution of energies among the molecules within each system. For systems that consist of independent molecules, such as an ideal gas, the relationship is

$$Q = q^N \qquad (11.5)$$

when the N particles in the system are distinguishable, and

$$Q = \frac{q^N}{N!} \qquad (11.6)$$

when the N particles in the system are indistinguishable.

11.3 OPEN SYSTEMS

Open systems are defined as systems that allow flow of both mass and energy across the boundaries that define its volume. They are characterized in terms of the three state functions μ, V, and T, where μ represents the chemical potential as defined in Section 6.5. As with the closed system, we need to consider an ensemble of systems where the members differ by both the number of molecules, N_i and the total energy, E_i as illustrated Figure 11.3.

In this case, we expect the joint probability of finding a member of the ensemble with energy E_i and number of molecules N_j will be the fraction of molecules n_{ij}/N that have those exact energies and numbers of molecules.

$$p(E_i, N_j) = \frac{n_{ij}}{N} = \frac{g_i e^{-\frac{E_i}{kT}} e^{\frac{N_j \mu}{kT}}}{\Xi}, \qquad (11.7)$$

where

$$\Xi = \sum_i \sum_j g_i e^{-\frac{E_i}{kT}} e^{\frac{N_j \mu}{kT}} = Q \sum_j e^{\frac{N_j \mu}{kT}}. \qquad (11.8)$$

The normalization here leads to the **grand canonical partition function** which now reflects both the distribution of energy and the distribution

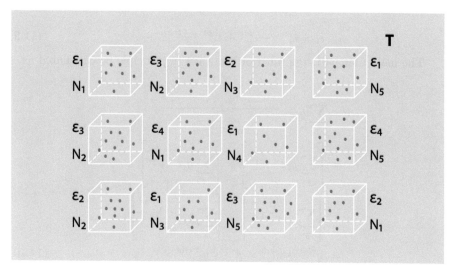

Figure 11.3: Illustration of an ensemble of open systems in a common thermal bath at temperature T with varying N and E.

of numbers of molecules across the systems in the ensemble. This latter distribution will become important later.

> Note that $N\mu$ has units of energy since the chemical potential is the free energy per mole.

11.4   RELATIONSHIPS BETWEEN PARTITION FUNCTIONS AND THERMODYNAMIC FUNCTIONS

The partition functions are particularly useful since they can be used to derive thermodynamic properties of materials from the molecular properties of the system. Calculation of the partition functions is then important. We saw that the grand canonical partition function (for open systems) depends linearly on the canonical partition function (for closed systems) (Equation 11.8 which in turn depends on the Nth power of the microcanonical partition function (for isolated systems) (Equations 11.5 and 11.6). The general problem then depends on calculating the molecular partition function, q.

In the most general terms, the energy of an individual molecule, ϵ_i is the sum of the energy associated with all of its electronic states, its vibrational states, its rotational states, and its translational states.[1]

---

[1] There may be additional energy terms depending on the state of the molecule and this formulation assumes that the energies are independent.

$$\epsilon_i = \epsilon_i^{elec} + \epsilon_i^{vib} + \epsilon_i^{rot} + \epsilon_i^{trans} + \ldots \tag{11.9}$$

The molecular partition function can then in principle be calculated as

$$
\begin{aligned}
q &= \sum_j g_j e^{-\frac{\epsilon_j^{elec} + \epsilon_j^{vib} + \epsilon_j^{rot} + \epsilon_j^{trans} + \ldots}{kT}} \\
&= \left[\sum_j g_j e^{-\frac{\epsilon_j^{elec}}{kT}} \right] \times \left[\sum_j g_j e^{-\frac{\epsilon_j^{vib}}{kT}} \right] \times \\
&\quad \left[\sum_j g_j e^{-\frac{\epsilon_j^{rot}}{kT}} \right] \times \left[\sum_j g_j e^{-\frac{\epsilon_j^{trans}}{kT}} \right] \times \ldots \\
&= q^{elec} \times q^{vib} \times q^{rot} \times q^{trans} \times \ldots \tag{11.10}
\end{aligned}
$$

We saw in Chapter 4 that we can use quantum mechanics to calculate the electronic, vibrational, and rotational energy states and the distribution of molecules among them as a function of temperature, so in principle we can calculate the electronic, vibrational and rotational microcanonical partition functions. Likewise, for a gas, the Maxwell–Boltzmann distribution can be used to calculate the translational microcanonical partition function.

The relationships between the partition functions and the thermodynamic state functions depend on the type of system.

For an **isolated system**, in which we define N, V, and E, we can calculate S, T, P, and μ from the microcanonical partition function or its derivatives:

$$S(N, V, E) = k \ln q$$

$$\frac{1}{T} = k \left(\frac{\partial \ln q}{\partial E} \right)_{VN}$$

$$P = kT \left(\frac{\partial \ln q}{\partial V} \right)_{EN}$$

$$\mu = -kT \left(\frac{\partial \ln q}{\partial N} \right)_{VE}. \tag{11.11}$$

For a **closed system**, in which we define N, V, and T, we can calculate A, S, P, μ, and E from the canonical partition functions or its derivatives

$$A(N, V, T) = kT ln Q$$

$$S = kT \left(\frac{\partial ln Q}{\partial T} \right)_{VN} + kln Q$$

$$P = kT \left(\frac{\partial ln Q}{\partial V} \right)_{TN}$$

$$\mu = -kT \left(\frac{\partial ln Q}{\partial N} \right)_{VT}$$

$$E = kT^2 \left(\frac{\partial ln Q}{\partial T} \right)_{VN}. \tag{11.12}$$

For an **open system**, in which we define V, T, and μ, we can calculate S, P, and N

$$PV = kln\Xi$$

$$S = kT \left(\frac{\partial ln\Xi}{\partial T} \right)_{V\mu}$$

$$P = kT \left(\frac{\partial ln\Xi}{\partial V} \right)_{T\mu} = \frac{kT ln\Xi}{V}$$

$$N = kT \left(\frac{\partial ln\Xi}{\partial \mu} \right)_{VT}. \tag{11.13}$$

As a consequence we have, a link between the distribution of energies of across individual molecules and the key thermodynamic properties, which is our link between molecular properties and bulk properties.

CHAPTER 12

Fluctuations in closed and open systems

12.1 THE CONCEPT OF FLUCTUATIONS

If we consider a single isolated system, the energy or the numbers of molecules in the system will never change. However, if we observe a closed system for a period of time, we expect the energy to vary with time; and if we observe an open system for a period of time, we expect the energy and the number of molecules to vary with time. In other words, these key system parameters will fluctuate with time.

If we monitor the fluctuations in energy and numbers of molecules with time and record the values, we will find that these values distribute in a manner equivalent to the distributions observed across the ensemble of systems we saw in Sections 11.2 and 11.3. The principle that the distribution of energy and numbers of molecules is the same whether we observe a large ensemble of equivalent systems or the same system over a long period of time is called the **ergodic hypothesis**. In other words, the distributions in Equations 11.3 and 11.7 hold for both temporal fluctuations and spatial fluctuations.

An absolute fluctuation δx is the difference between any given value of x and the mean value of the distribution $\langle x \rangle$.[1] so $\delta x = x - \langle x \rangle$. The average fluctuation, $\langle \delta x^2 \rangle^{0.5} = \langle (x - \langle x \rangle)^2 \rangle^{0.5}$, is the standard deviation of the absolute fluctuation. The relative fluctuation is the fluctuation relative to the mean, which means that the average relative fluctuation is then described through the ratio of the standard deviation and the mean or the variance and the square of the mean

$$\left(\frac{\langle \delta x \rangle}{\langle x \rangle} \right) = \left(\frac{\langle (x - \langle x \rangle)^2 \rangle}{\langle x \rangle^2} \right)^{0.5} = \left(\frac{\sigma^2}{\mu^2} \right)^{0.5}. \tag{12.1}$$

---

[1]We will use the $\langle x \rangle$ symbols to indicate the average over space (the ensemble average) and the symbol $\bar{x}$ to indicate an average over time.

Note that the variance $\sigma^2 = \langle (x - \langle x \rangle)^2 \rangle$ can be rewritten as $\sigma^2 = \langle x^2 \rangle - \langle x \rangle^2$ (see problem assignment).

This average relative fluctuation is well defined for the binomial and Poisson distributions and both vary as the inverse of the square root of the mean value

$$\left(\frac{\sigma^2}{\mu^2} \right)^{0.5} = \left(\frac{1-p}{\mu} \right)^{0.5} - binomial$$

$$\left(\frac{\sigma^2}{\mu^2} \right)^{0.5} = \left(\frac{1}{\mu} \right)^{0.5} - Poisson.$$

We also note that the exponent of the normal distribution is a measure of the fluctuation relative to the variance.

12.2 FLUCTUATIONS OF ENERGY IN CLOSED SYSTEMS

In the closed system, the energy is free to fluctuate about some mean value, $\langle E \rangle$. The energy distribution is given by Equation 11.3 and is related to the Canonical partition function, Q, so the mean energy can be calculated as

$$\langle E \rangle = \sum_i E_i p(E_i) = \sum_i E_i \frac{g_i e^{-E_i/kT}}{Q}. \tag{12.2}$$

Correspondingly, the second moment of the distribution of energy can be calculated as

$$\langle E^2 \rangle = \sum_i E_i^2 p(E_i) = \sum_i E_i^2 \frac{g_i e^{-E_i/kT}}{Q}. \tag{12.3}$$

Recall from Section 11.4 that in a closed system, the energy is related to the derivative of the logarithm of the Canonical partition function. Knowing that the energy can fluctuate, we now more correctly indicate that this is the average energy of the system, i.e.,

$$\langle E \rangle = kT^2 \left(\frac{\partial lnQ}{\partial T} \right)_{VN} \tag{12.4}$$

We can differentiate this with respect to temperature (see problem assignment) to show that

$$\left(\frac{\partial \langle E \rangle}{\partial T} \right)_{VN} = \frac{1}{kT^2} \sigma_E^2. \tag{12.5}$$

But the temperature derivative of the energy at constant volume is the heat capacity at constant volume, C_V (See Equation 6.11), so Equation 12.3 can be rewritten as

$$\sigma_E^2 = kT^2 C_V. \tag{12.6}$$

For an ideal gas, we expect the energy to be $\langle E \rangle = \frac{3}{2}NkT$ and therefore $C_V = \frac{3}{2}Nk$, so for an ideal gas the variance of the energy fluctuations relative to the square of the mean energy is

$$\left(\frac{\sigma_E^2}{\langle E \rangle^2} \right) = \frac{kT^2 \left(\frac{3}{2}Nk \right)}{\left(\frac{3}{2}NkT \right)^2} = \frac{2}{3}\frac{1}{N}. \tag{12.7}$$

This gives us the remarkable result that the relative fluctuation in energy in a closed system is proportional to the inverse of the square root of the number of molecules in the system. If we consider macroscopic systems, where N may be on the order of Avogadro's number, the relative fluctuation in energy is expected to be very small. For example, for a system with micromolar concentrations, the total number of molecules in volume of a liter is on the order of 6×10^{17} and the relative fluctuation is therefore on the order of 1.3×10^{-9}. We are therefore indeed justified in proposing that the mean energy is the same as the peak energy in a closed system, which is what was needed to get to the relationships in Section 11.4.

Even in systems other than an ideal gas, we expect the mean energy to be on the order of NkT and the heat capacity to be on the order of Nk and hence the relative fluctuation will still be on the order of $\frac{1}{\sqrt{N}}$

We can surmise that as the system gets smaller, there are fewer molecules and hence the relative fluctuations in energy could become important. For example, as per Table 7.2, nanoparticles of diameters less than about 6 nm have fewer than about 10^4 molecules and the energy fluctuations would be greater than 1 % and is perhaps as much as 10 % for the surface molecules on a 2 nm particle.

12.3   DISTRIBUTION OF ENERGY AMONG INDIVIDUAL MOLECULES

If we consider the mean total energy as the sum of the energy of the individual molecules, we expect that $\langle E \rangle = N \langle \epsilon \rangle$, which implies that $\sigma_\epsilon^2 = kT^2 C_V / N$, and hence

$$\left(\frac{\sigma_\epsilon^2}{\langle \epsilon \rangle^2} \right) \approx \frac{kT^2 (Nk)}{(kT)^2 N} = 1. \tag{12.8}$$

It is evident that the fluctuation of energy for individual molecules can be very large, in concert with the observation that the energy distribution for a collection of molecules in an isolated system is very broad, as exemplified by the Maxwell–Boltzmann velocity distribution shown in Figure 10.9.

While the distribution of energy among individual molecules is broad, the distribution of total energy for the collection of molecules is narrow except for very small systems.

> Note that the narrow distribution of the total energy allows us to substitute the mean energy by the most probable energy (the mode), which is one of the underlying approximations that led to the results of Section 11.4.

12.4 FLUCTUATIONS OF ENERGY AND NUMBERS IN OPEN SYSTEMS

In an open system, both the energy and the number of molecules can fluctuate. The relative energy fluctuations will depend on the inverse of the square root of the number of molecules as in the closed systems which was shown above in Equation 12.5.

The average number of molecules, $\langle N \rangle$, can be calculated using the probability distribution in Equation 11.7 and hence

$$\langle N \rangle = \sum_i N_i p(E_i, N_j) = \sum_i N_i \frac{g_i e^{-\frac{E_i}{kT}} e^{\frac{N_j \mu}{kT}}}{\Xi}. \tag{12.9}$$

Correspondingly, the second moment of the distribution of numbers of molecules can be calculated as

$$\langle N^2 \rangle = \sum_j N_j^2 p(E_i, N_j) = \sum_j N_j^2 \frac{g_i e^{-\frac{E_i}{kT}} e^{\frac{N_j \mu}{kT}}}{\Xi}. \tag{12.10}$$

We also note from Section 11.4 that the number of molecules in an open system is related to the derivative of the Grand Canonical partition function with respect to the chemical potential. If we identify this with the average number of molecules we get

$$\langle N \rangle = kT \left(\frac{\partial ln\Xi}{\partial \mu} \right)_{VT}. \tag{12.11}$$

Differentiating with respect to the chemical potential gives us (see problem assignment) in analogy with the procedure for the energy above, that

$$\left(\frac{\partial \langle N \rangle}{\partial \mu} \right)_{VT} = \frac{1}{kT} \sigma_N^2. \tag{12.12}$$

The derivative of $\langle N \rangle$ with respect to μ can be shown (see problem assignment) to be proportional to the isothermal compressibility, κ_T, the square of the average number of molecules, and the inverse of the volume and as a result

$$\sigma_N^2 = \frac{kT \langle N \rangle^2 \kappa_T}{V}, \tag{12.13}$$

where we recall from Table 6.2 that $\kappa_T = -\frac{1}{V} \left(\frac{\partial V}{\partial P} \right)_{TN}$.

For an ideal gas, $\kappa_T = \frac{1}{P} = \frac{V}{\langle N \rangle kT}$, so

$$\left(\frac{\sigma_N^2}{\langle N \rangle^2} \right) = \frac{kT \langle N \rangle^2}{V} \frac{V}{\langle N \rangle \, kT} \frac{1}{\langle N \rangle^2} = \frac{1}{\langle N \rangle}. \tag{12.14}$$

We see once again that the relative fluctuation is proportional to $\frac{1}{\sqrt{\langle N \rangle}}$. This is very small for most macroscopic systems, but can be expected to be significant as the numbers of molecules in the system of interest decreases.

The significance of the above discussion is that *fluctuations* in energy or in occupation number can be directly related to the *average* occupation number in open systems. This opens up the opportunity to measure fluctuations as a means to determine concentrations, as will be discussed in more detail in the following chapter.

CHAPTER 13

Measuring concentration changes

The previous chapter demonstrates that the number of molecules in a fixed volume will fluctuate even when the system is at equilibrium. This fluctuation for a single molecule will be a consequence of movement in and out of the volume of interest which implies that the time course of the fluctuation may reveal information about the transport dynamics.

Correspondingly, if there are two molecules that can react to form a third, as in the simple chemical reaction

$$A + B = C,$$

then we may expect that the numbers of molecules of A, B, and C will fluctuate both because of the transport dynamics and because there is a constant interconversion even at equilibrium. This implies that the time course of the fluctuations may reveal information about both transport dynamics and the kinetics of the reaction.

The arguments in the previous chapter show us that if the concentrations or volumes are large, the relative fluctuations will be too small to detect, but in nanoscale systems, it is, in principle, possible to measure these concentration fluctuations. The purpose of this chapter is to explore how analysis of fluctuations can be explored to measure dynamic processes in small systems.

13.1   THE CHALLENGE OF MEASURING DYNAMIC PROCESSES

We will consider two types of dynamic processes of interest: transport (by diffusion or flow) and kinetics of chemical reactions.

13.1.1 Transport phenomena

Molecules in a gas will move at a constant speed in a particular direction until it encounters another molecule and collides with it. In the collision there may be a momentum transfer which leads to changes in speed and direction of movement for both molecules.

From the Maxwell–Boltzmann distribution of speeds (Equation 10.13) it is possible to calculate the mean speed as

$$\langle v \rangle = \sqrt{\left(\frac{8kT}{\pi m} \right)}, \tag{13.1}$$

where m is the mass of the molecule of interest.

If the molecule has a radius of r it will collide with any molecule within a distance of $2r$ from its center, so within an area of $\sigma = 4\pi r^2$, which is termed the collision cross-section for the molecule. For a gas at a density of $\rho = \langle N \rangle /V$, the frequency of collisions of that single molecule then becomes

$$z_{coll} = \sigma \rho \langle v \rangle = \sigma \rho \sqrt{\left(\frac{8kT}{\pi \mu} \right)}, \tag{13.2}$$

where we have substituted the mass m by the reduced mass μ to account for the fact that all the molecules are moving so that we need to use the relative speed (see for example McQuarrie or Atkins for details).

To get the collision frequency for all the molecules, we multiply by the total number of molecules, $\langle N \rangle$, and divide by 2 to avoid double counting collisions. This then leads to the collision frequency per unit volume (by dividing by the volume)

$$Z_{AA} = \frac{\frac{1}{2} \langle N \rangle z_{coll}}{V} = \sigma \sqrt{\left(\frac{2kT}{\pi \mu} \right)} \frac{\langle N \rangle^2}{V^2}. \tag{13.3}$$

More generally, if we consider collisions of different molecules, as needed in a chemical reaction, we obtain

$$Z_{AB} = \sigma_{AB} \sqrt{\left(\frac{8kT}{\pi \mu} \right)} \frac{\langle N_A \rangle \langle N_B \rangle}{V^2}, \tag{13.4}$$

where $\sigma_{AB} = \pi(r_A + r_B)$ is the collision cross section for the two molecules, A and B.

If the collision frequency is z_{coll}, then the average time between collisions for a single molecule will be $\frac{1}{z_{coll}}$ and in that time, the average distance traveled by the molecule will be $\lambda = \frac{\langle v \rangle}{z_{coll}}$. This is called the mean free path for the molecule.

In liquids, the molecules will collide continually and the mean free path is on the order of the dimension of the molecule or less.

13.1.1.1 *Fick's Laws and the diffusion coefficient*

Consider the motion of molecules in a system where there is a gradient in concentration. In that case, we expect the frequency of collisions will be greater and hence the mean free path will be smaller at the higher concentrations while the frequency of collisions will be smaller and the mean free path will be longer at the lower concentrations. The net effect is that there will be a flux, J_x, of molecules in the direction from higher to lower concentrations. By Fick's first law the flux will be proportional to the gradient in concentration, so that

$$J_x = D\frac{\partial N}{\partial x}, \tag{13.5}$$

where the proportionality constant, D, is the diffusion coefficient.

For an ideal gas, the diffusion coefficient is related to the product of the mean free path and the mean velocity

$$D = \frac{1}{3}\lambda \langle v \rangle . \tag{13.6}$$

Note that the units of the diffusion coefficient is distance squared per unit time for example m^2/s.

Expressions for the diffusion coefficient can be obtained if we consider what will happen when a large number of molecules are initially confined to a thin slab in the x, y-plane. With time, the molecules will move along the z-direction into the space on either side of the slab in a manner that satisfies the diffusion equation or Fick's second law, which can be derived from Fick's first law (See problem assignments)

$$\left(\frac{\partial N(z;t)}{\partial t}\right) = D\left(\frac{\partial^2 N(z;t)}{\partial z^2}\right) \tag{13.7}$$

This partial differential equation can be solved to show that the distribution of molecules as a function of position along the z-axis and as a function of time becomes

$$N(z;t) = \frac{N_0}{A\sqrt{\pi Dt}}e^{-\frac{z^2}{4Dt}}, \tag{13.8}$$

where A is the area of the slab.

We recognize that at any fixed time, t, this distribution looks like the normal distribution (see Section 10.1.3.3). The first moment with respect to the z-coordinate of this distribution is zero, which means that the molecules will move equally well in the positive and negative direction. The second moment with respect to the z-coordinate of this distribution is

$$\langle z^2 \rangle = 2Dt. \tag{13.9}$$

This provides an interpretation of the diffusion coefficient in terms of the mean square distance traveled by the molecules at a particular time.

The expression in Equation 13.9 is obtained for a one-dimensional diffusion problem and it can be shown that in general, the mean square displacement is

$$\langle r^2 \rangle = 2dDt, \tag{13.10}$$

where d is the order of dimension of the space in which diffusion takes place, i.e., $d = 1$ for 1D-diffusion, $d = 2$ for 2D-diffusion, and $d = 3$ for 3D-diffusion.

This is also known as the Einstein equation for diffusion which was developed in 1905. Einstein also showed that the diffusion coefficient is related to the terminal velocity, v_t, of a particle moving in response to an applied force, F, as

$$D = kT\frac{v_t}{F} = \frac{kT}{f}, \tag{13.11}$$

where f is the frictional coefficient for the particle in the medium in which it is moving.

Stokes showed that the frictional coefficient for a sphere of radius, r, moving in a medium of viscosity, η, is $f = 6\pi\eta r$, so in that case $D = \frac{kT}{6\pi\eta r}$. This suggests that a larger particle will have a smaller diffusion coefficient and hence a smaller mean square displacement at any given time.

Note that if a particle is flowing, the velocity, v, determines the displacement as a function of time, and in comparison with diffusion we would see that the mean square displacement is a function of time squared

$$\langle r^2 \rangle = v^2 t^2. \tag{13.12}$$

One way to measure the diffusion coefficient or other transport mechanisms is to create a concentration gradient and then measure how that gradient changes with time. This is possible in many systems and is the classical way to measure transport phenomena.

We shall see later that it is also possible to measure the rates of diffusion or flow of molecules by measuring fluctuations of numbers of molecules in small volumes.

13.1.2 Reaction kinetics

Chemical reactions will change the concentrations of the species involved in the reaction until an equilibrium is reached, but even at equilibrium, the concentrations will fluctuate. Recall that we describe reactions in terms of the order of reactions, which reflects the number of species involved in the rate limiting step of the reaction.

We are generally interested in measuring the rate of a chemical reaction because it can be a clue to the mechanism of the reaction. In order to measure the rate, we would initiate the reaction with only the reactant(s) present and then measure its (their) disappearance or the appearance of the product(s).

First order kinetics For a first order reaction $A \overset{k_1}{\rightarrow} B$[1] the rate of disappearance of the reactant is

$$-\frac{d[A]}{dt} = k_1[A],\qquad(13.13)$$

which leads to

$$[A] = [A]_0 e^{-k_1 t},\qquad(13.14)$$

where $[A]_0$ is the initial concentration of the reactant.

Correspondingly, the rate of appearance of the product (see problems assignment) is

$$[B] = [A]_0\left((1 - e^{-k_1 t})\right).\qquad(13.15)$$

When the reaction is reversible, the reverse reaction $B \overset{k_2}{\rightarrow} A$ will become important as the product concentration increases and the reactant concentration decreases. At some point in time, the concentrations reach the equilibrium concentrations where the rate of the forward reaction equals that of the reverse reaction, and hence

$$-\frac{d[A]}{dt} = k_1[A]_{eq} = \frac{d[B]}{dt} = k_2[B]_{eq},\qquad(13.16)$$

which tells us that the equilibrium constant $K = \frac{[B]_{eq}}{[A]_{eq}} = \frac{k_1}{k_2}$.

Parallel first order kinetics We shall later encounter parallel first order reactions, in which a single reactant can yield two or more products: $A \overset{k_B}{\rightarrow} B$ and $A \overset{k_C}{\rightarrow} C$. This leads to a faster rate of disappearance of A such that

$$-\frac{d[A]}{dt} = (k_B + k_C)[A],\qquad(13.17)$$

which leads to

$$[A] = [A]_0 e^{-(k_B+k_C)t}.\qquad(13.18)$$

Correspondingly,

$$\frac{d[B]}{dt} = k_B[A] = k_B[A]_0 e^{-(k_B+k_C)t}$$
$$\frac{d[C]}{dt} = k_C[A] = k_C[A]_0 e^{-(k_B+k_C)t}\qquad(13.19)$$

---

[1] We will use the symbol $\rightarrow$ for the forward reaction, $\leftarrow$ for the reverse reaction, $\rightleftharpoons$ for a reversible reaction, and $=$ for the reaction at equilibrium.

and

$$[B] = k_B \frac{[A]_0}{(k_B + k_C)} \left(1 - e^{-(k_B + k_C)t} \right)$$

$$[C] = k_C \frac{[A]_0}{(k_B + k_C)} \left(1 - e^{-(k_B + k_C)t} \right). \tag{13.20}$$

It follows that the two products appear at the same rate and in a constant ratio determined only by their relative rate constants $\frac{[B]}{[C]} = \frac{k_B}{k_C}$. We can express this in terms of the yield of each of these products relative to the total product formation as

$$\frac{[B]}{[B] + [C]} = \frac{k_B}{k_B + k_C}$$

$$\frac{[C]}{[B] + [C]} = \frac{k_C}{k_B + k_C}. \tag{13.21}$$

We can generalize this to multiple parallel reactions $A \xrightarrow{k_i} P_i$ to get

$$[A] = [A]_0 e^{-\left(\sum_i k_i \right) t} \tag{13.22}$$

$$\frac{[P_i]}{\sum_i [P_i]} = \frac{k_i}{\sum_i k_i}. \tag{13.23}$$

So the larger the number of parallel reactions, the faster the decay of the reactant and the lower the yield of any one of the products.

Second order kinetics   The second order reaction involves two reactants combining to form one or more products. Let us consider two simple examples, that of identical reactants and that of different reactants.

The reaction $A + A \xrightarrow{k_1} P$ has the rate equation

$$-\frac{d[A]}{dt} = k_1 [A]^2, \tag{13.24}$$

which leads to

$$\frac{1}{[A]} = kt + \frac{1}{[A]_0}. \tag{13.25}$$

The reaction $A + B \xrightarrow{k_1} C$ could be first order in each of the reactants and second order overall. This reaction would become pseudo-first order overall if one of the reactants is in excess so that the concentration remains nearly constant during the reaction.

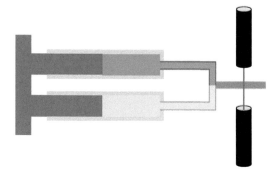

Figure 13.1: Illustration of a syringe mixing system.

The general solution, when the initial concentrations of A and B are different, is

$$ln\left(\frac{[A]_0[B]}{[B]_0[A]}\right) = ([B]_0 - [A]_0)\, kt \tag{13.26}$$

and when the initial concentrations are the same $[B]_0 = [A]_0$, the problem reduces to that of Equation 13.24 with the solution given by Equation 13.25.

In all cases, it is possible to determine the order of the reaction by determining the concentrations of the reactants or the products and plotting them as a function of time and then fit them to one of the functions provided above. Thus, we are faced with determining concentrations as a function of time.

13.1.3 Measuring concentrations

The tools available for measuring concentrations of reactants or products of chemical reactions depend on the speed with which the reactions proceed.

For slow reactions (minutes or slower) it is possible to simply mix the reactants to initiate the reaction and follow some indicator of the reaction, such as the disappearance or appearance of a spectroscopic signature (absorbance, fluorescence, etc.). The limitations are the mixing time and the sensitivity and speed of the detection tool relative to the reaction time.

For faster reactions in which the mixing time is the limiting feature, the speed of mixing can be enhanced by techniques such as a stopped-flow apparatus, in which two syringes are rapidly pushed and the content are caused to rapidly mix in a chamber as illustrated in Figure 13.1. Through these mixing techniques, reactions that happen in the millisecond regime or slower can now be measured.

To overcome the time limitations of mixing, it is possible to apply perturbations techniques. The principle is to monitor a system at equilibrium and then perturb that equilibrium by changing a parameter, such as temperature

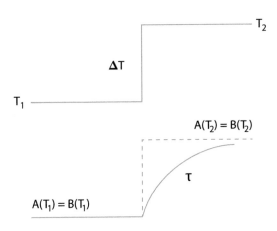

Figure 13.2: Illustration of the principle of a perturbation relaxation experiment.

or pressure, that will change the equilibrium constant for the reaction and therefore shift the equilibrium. By observing the changes in concentrations as the system approaches the new equilibrium, one can measure the rate of change, $1/\tau$ which is the sum of the rate constants for the forward, k_1, and reverse, k_2, reactions.

$$\frac{1}{\tau} = k_1 + k_2 \tag{13.27}$$

Since the equilibrium constant is also related to these rate constants as

$$K = \frac{k_1}{k_2}, \tag{13.28}$$

we can determine the rate constants from these two equations with two unknowns.

The principle of the perturbation approach is illustrated in Figure 13.2 for the case of a temperature jump for a general reaction $A = B$ at its equilibrium.

Temperature jumps   The equilibrium constant depends on temperature whenever the enthalpy of the reactants differs from the enthalpy of the products so that there is a net enthalpy of reactions, ΔH_r^o. In that case, the temperature dependence of the equilibrium constant is given by the van't Hoff equation

$$\frac{dlnK}{dT} = \frac{\Delta H_r^o}{RT^2}. \tag{13.29}$$

If the volume of observation is small, then the temperature changes can be very rapid and it is possible to measure the time course of reactions on the microsecond time scale.

In general, it is easier to raise the temperature of a solution by heating, than it is to cool it, and hence it may take time to return to the original temperature. However, it is still possible to make the perturbation of short duration so that the temperature change is small which means that the experiment can be repeated many times to provide opportunities for signal averaging and hence better precision in the measurement.

Pressure perturbations The equilibrium constant depends on the pressure whenever the molar volume of the reactants differs from the molar volume of the products so that there is a net volume of reaction V_r^o. In that case,

$$\frac{dlnK}{dT} = -\frac{V_r^o}{RT}. \tag{13.30}$$

Using piezoelectric systems, the pressure perturbations can be very rapid, can produce both positive and negative changes in pressure, and can be repeated rapidly, which is often needed for signal averaging since the volume of reaction can be quite small.

Other perturbation techniques In principle, any parameter that will alter the chemical equilibrium can be used as a perturbation source. For example, if the reaction leads to changes in the number of charges or a change in the dielectric constant of the solution, introduction of an electric field, E, may shift the equilibrium. In that case,

$$\frac{dlnK}{dE} = \frac{\Delta \epsilon V E}{RT}. \tag{13.31}$$

Correspondingly, if the reaction either uses or produces hydrogen ions, a pH change will shift the equilibrium. In this latter case, mixing may, however, limit the speed of reactions that can be measured and it may be necessary to use tools such as the stopped-flow approaches even though changes in hydrogen ion concentration across a solution are very rapid.

13.2 FLUCTUATIONS AND CHEMICAL KINETICS

We recognized earlier that in open systems the properties of the system can fluctuate. In principle, nature provides through local fluctuations in temperature and density its own local perturbations. If we observe systems in small, fixed volumes, the numbers of molecules in those systems will be small and the fluctuations in concentration may be measurable. Since the magnitude of the fluctuations vary as the inverse of the square root of the number of molecules, we can establish the precision we need to make a meaningful measurement.

Table 13.1: Sensitivity (signal-to-noise) Needed to Measure Fluctuations for a Given Total Number of Molecules

N	Sensitivity	Molarity (pL)	Molarity (fL)
10^8	1 in 10^4	160 μM	160 mM
10^6	1 in 10^3	1.6 μM	1.6 mM
10^4	1 in 10^2	16 nM	16 μM
10^2	1 in 10^1	160 pM	160 nM
1	1 in 1	1.6 pM	1.6 nM

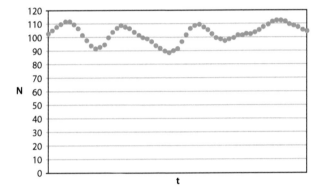

Figure 13.3: Variation in the number of molecules, N as a function of time.

Table 13.1 illustrates the sensitivity (or signal to noise) needed to measure a fluctuation for a series of different total number of molecules and the corresponding concentrations needed if these molecules are in a pL and fL volume, respectively.

our ability to measure concentrations more precisely improves, we can measure the results of spontaneous number fluctuations in small volumes at higher concentrations and from these fluctuations extract dynamic properties such as the transport rates or chemical reaction rates as well as the actual average concentration.

13.2.1 Concentration correlation spectroscopy

The concept of concentrations correlation spectroscopy was first developed and experimentally realized in the early 1970s and an excellent summary was published by Elson and Webb in 1975 (Ann. Rev. Biophys. Bioeng. 4, 311-334).

Consider measurements of the number of molecules in a given open volume as a function of time as illustrated in Figure 13.3. The number in this example fluctuates around a mean value of 103 with a standard deviation of about 10.

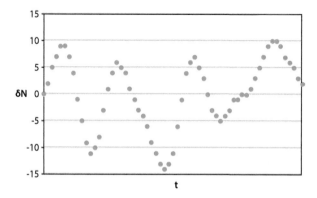

Figure 13.4: Fluctuation $\delta N = N - \langle N \rangle$ in the number of molecules as a function of time.

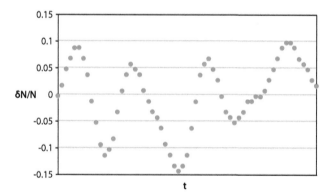

Figure 13.5: Relative fluctuation $\delta N / \langle N \rangle = (N - \langle N \rangle) / \langle N \rangle$ in the number of molecules as a function of time.

We can calculate the fluctuations about the mean value $\delta N = N - \langle N \rangle$ corresponding to these measurements as illustrated in Figure 13.4. Figure 13.5 shows the corresponding relative fluctuations $\delta N / \langle N \rangle = (N - \langle N \rangle) / \langle N \rangle$.

It is clear that these fluctuations vary in their amplitudes as the numbers of molecules randomly move in and out of the observation volume. Likewise, the fluctuations have a specific duration which represents the underlying dynamic process.

The mathematical tool used to analyze the fluctuations is the correlation function.

13.2.1.1 Correlation functions and their interpretation

There are several forms of the correlation function depending on whether they are calculated directly from the numbers shown in Figure 13.3, from the fluctuations shown in Figure 13.4, or from the relative fluctuations shown in Figure 13.5.

The number correlation function is calculated by

$$G_N(j\Delta t) = \frac{1}{k-j} \sum_{i=1}^{k-j} N(t_i) \times N(t_{i+j}). \tag{13.32}$$

Here we have introduced several new concepts. First, Δt is the time interval between individual measurements in the time sequence, indicated by the vertical lines in the three figures. Second, k is the total number of measurements in the time sequence. Third, i and j are the markers for the i-th and j-th time interval, respectively, so that t_i and t_{i+j} are separated in time by $j\Delta t$ time intervals.

The correlation function at a particular value of j, i.e., for a particular separation in time, is then the sum of all of the pairwise products of the values of N separated by that separation of time.

Thus, if $j = 1$, then the correlation function becomes

$$G_N(1\Delta t) = \frac{1}{k-1} (N_1 \times N_2 + N_2 \times N_3 + N_3 \times N_4 + N_4 \times N_5 \dots), \tag{13.33}$$

where the sum is over $k - 1$ data points.

Correspondingly, if $j = 2$

$$G_N(2\Delta t) = \frac{1}{k-2} (N_1 \times N_3 + N_2 \times N_4 + N_3 \times N_5 + N_4 \times N_6 \dots), \tag{13.34}$$

where the sum is over $k - 2$ data points.

This way one can build up a correlation function which is a function of the difference in time between the points that are being calculated.

The correlation function calculated from the fluctuations is

$$G(j\Delta t) = \frac{1}{k-j} \sum_{i=1}^{k-j} (N(t_i) - \langle N \rangle) \times (N(t_{i+j}) - \langle N \rangle). \tag{13.35}$$

Since we can show (see problem assignment) that

$$\frac{1}{k-j} \sum_{i=1}^{k-j} (N(t_i) - \langle N \rangle) \times (N(t_{i+j}) - \langle N \rangle) =$$

$$\frac{1}{k-j} \sum_{i=1}^{k-j} N(t_i) \times N(t_{i+j}) - \langle N \rangle^2 \tag{13.36}$$

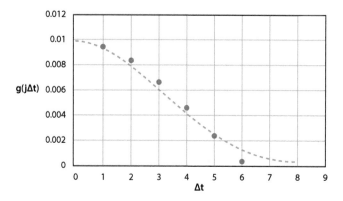

Figure 13.6: The correlation function corresponding to the data in Figure 13.5 and the fit to a Gaussian decay.

we see that the number correlation function is related to the fluctuation correlation function by the square of the mean.

$$G(j\Delta t) = G_N(j\Delta t) - \langle N \rangle^2 \tag{13.37}$$

The correlation function calculated from the relative fluctuations is in turn

$$
\begin{aligned}
g(j\Delta t) &= \frac{\frac{1}{k-j}\sum_{i=1}^{k-j}\left(N(t_i) - \langle N \rangle\right) \times \left(N(t_{i+j}) - \langle N \rangle\right)}{\langle N \rangle^2} \\
&= \frac{G(j\Delta t)}{\langle N \rangle^2} \\
&= \frac{G_N(j\Delta t)}{\langle N \rangle^2} - 1.
\end{aligned}
\tag{13.38}
$$

This latter correlation function is shown for $j = 1 - 6$ in Figure 13.6.

The correlation function derives its name from the fact that it determines what correlations, if any, exist between different portions of the time series of measurements. When two time points are close, we expect that they represent the result of the same fluctuation and hence the value of the relative fluctuations will have the same sign and the products will mostly be positive so the sum will be positive. However, when two time points are far apart, they are likely to represent different fluctuations and the value of the relative fluctuation will randomly have the same or different signs and the product will randomly be positive or negative and therefore the sum will average to zero. This feature is seen in Figure 13.6 where the first few points are positive and approach zero as $j \to 6$.

The correlation function has three features: an amplitude as $j \to 0$; a characteristic decay shape; and a characteristic decay time.

The correlation function amplitude   The amplitude of the correlation function arises when j approaches zero. We note that this corresponds to

$$g(0) = \frac{\frac{1}{k}\sum_{i=1}^{k}(N(t_i) - \langle N\rangle) \times (N(t_i) - \langle N\rangle)}{\langle N\rangle^2} = \frac{\sigma_N^2}{\langle N\rangle^2}, \qquad (13.39)$$

so the amplitude of the correlation function calculated from the relative fluctuations is identically equal to the ratio of the variance and the mean squared. This means that the amplitude of the correlation function can be used to calculate the mean number of molecules in a volume as per Equation 12.14.

$$g(0) = \frac{1}{\langle N\rangle} \qquad (13.40)$$

While this is not interesting if we can directly measure the number of molecules, as in the example used above, it will be important, as we shall see later, when the concentration is measured through another parameter, such as the absorption or the emission of light.

The correlation function decay shape   The decay of the correlation function will depend on the origin of the fluctuations and the shape will therefore provide information about the underlying dynamic process or processes. For example, it is well known that processes that give rise to binary changes (off-on) as in a Markov process (see Section 10.1.3.2) will have a correlation function that decays exponentially. Thus we might expect a chemical reaction that fluctuates at equilibrium to lead to an exponentially decaying correlation function.

The decay illustrated in Figure 13.6 has been approximated by a decreasing function shaped as a Gaussian[2] which could reflect a flow of molecules through a laser beam with a Gaussian intensity profile (see later).

The correlation function decay time   The characteristic time for the decay of the correlation function is related to the rate constants for the dynamic process. For example, if the fluctuations arise because of a change in the equilibrium reaction, the characteristic time for the decay, or the correlation time, would reflect the sum of the forward and reverse reactions as shown in Equation 13.27.

The decay illustrated in Figure 13.6 has a decay time of three time units which, in this case, would reflect the speed with which the molecules flow through the laser beam.

---

[2]In this example the fluctuations in Figure 13.3 were created by superimposing a number of normal distributions with a mean of zero and a variance of 3.

Other properties of the correlation function   The correlation function is a useful device since it contains information about the origin of fluctuations in any system of interest. If the dynamic processes is a stationary process,[3] then the measurement of the correlation function is independent of when the measurement is taken and

$$G(\tau) = \langle \delta c(t)\delta c(t+\tau)\rangle = \langle \delta c(0)\delta c(\tau)\rangle,\qquad (13.41)$$

where we have used τ to represent the time interval difference, $j\Delta t$, so that the general form of the correlation function can be a continuous function.

> A stationary process is one in which the mean value is independent of time.

While it is always possible to calculate the correlation function directly as indicated in Equations 13.33 and 13.34, this calculation can be tedious and time consuming, even on modern computers. An alternative is to use the Wiener–Khinchin theorem, which tells us that the correlation function is the Fourier transform of the power spectrum, which can in turn be calculated from the Fourier Transform of the time series of interest. This principle is illustrated in Figure 13.7 in which the Fourier transform of the time series $c(t)$ gives a function, $C(\omega)$, with real and imaginary components. The power spectrum is the square of this Fourier transform, i.e., the product of the function and its complex conjugate $P(\omega) = C(\omega) \times C^*(\omega)$. Finally, the correlation function is the reverse Fourier Transform of the power spectrum.

Note that the Fourier transforms can work in both directions so that the original function, $c(t)$, can be recovered by the reverse Fourier transform of $C(\omega)$ or the power spectrum, $P(\omega)$, can be recovered by the forward transform of the correlation function, $G(\tau)$. However, the original function, $c(t)$ cannot be recovered from the correlation function $G(\tau)$, nor can its Fourier transform, $C(\omega)$ be recovered from the power spectrum, $P(\omega)$.

> The Fourier transform of a continuous time series, $c(t)$ is calculated as
>
> $$C(\omega) = \int_{-\infty}^{\infty} c(t)e^{-i\omega t}dt. \qquad (13.42)$$
>
> The reverse Fourier transform gives the same function back
>
> $$c(t) = \int_{\infty}^{\infty} C(\omega)e^{i\omega t}d\omega. \qquad (13.43)$$

[3]A stationary process is one that is independent of when in time it is being observed so, for example, the mean does not change with time, etc.

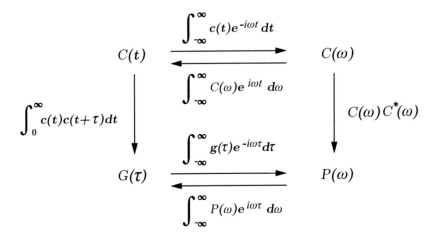

Figure 13.7: Relationships between the time series, $c(t)$, the power spectrum, $P(\omega)$, and the correlation function, $G(\tau)$.

13.2.1.2 Measures of concentration

It is rare that we can measure the numbers of molecules in our system directly, but there are a number of measurements that are linearly dependent on the concentration, and hence are suitable alternatives. Examples include absorption and emission of light.

Absorption of light As we saw in Part I, electrons that are confined in bonds or in a nanoparticles can be promoted from one quantum state to another provided the appropriate energy is provided. This energy can be provided in the form of electromagnetic radiation, which leads to the absorption of that radiation. As per the Beer–Lambert Law, the absorbance, A (or optical density, OD), is linearly proportional to the concentration, c, the molar absorption coefficient, ϵ (which is a reflection of the efficiency of absorption per molecule), and the dimension of the system, l (the length of the path):

$$A(t) = OD = log\left(\frac{I_0(t)}{I(t)}\right) = \epsilon l c(t), \tag{13.44}$$

where $I_0(t)$ is the intensity of the light incident on the sample and $I(t)$ is the intensity of light emerging from the sample (i.e., the difference between these quantities correspond to the intensity of light absorbed in the sample).

For most molecules, the molar absorption coefficient[4] is too small to make absorption measurements sensitive at low concentrations. For example, a

---

[4] or (molar) extinction coefficient

highly absorbing molecule may have an extinction coefficient on the order of $\epsilon = 10^6 \ L \ mol^{-1} \ cm^{-1}$ which implies that if we can measure the absorbance with a precision of 10^{-3} we can measure concentrations on the order of $10^{-9} \ M$ as long as the path length is 1 cm. For much smaller volumes where the path length may be in the order of 1 μm, the concentration needs to be closer to $10^{-5} \ M$ which may not lead to sufficiently large fluctuations to be measurable. Generally, absorption will not be sensitive enough, particularly since it is measured as a ratio of the input and output intensities.

Emission of light   The emission of light, as luminescence or fluorescence, depends on the number of molecules that absorb light that in turn emit that light. The intensity of emission is therefore proportional to the intensity of light absorbed, $I_0(t) - I(t)$, the fraction of excited molecules that emit light (also known as the quantum yield of fluorescence, Φ_F), and a proportionality constant, q, which depends on factors such as the efficiency of collection and detections of the photons emitted from the sample.

While absorption of light is generally from light traveling along a particular direction, the emission of light will occur in all directions (unless the light is polarized). It is therefore difficult to capture all of the emitted photons. Likewise, the detectors will not necessarily detect all the photons. The proportionality constant, q, takes these factors into account and will be a constant for a particular instrument and particular wavelengths. However, q will differ from instrument to instrument and for different wavelengths, so it is difficult to compare absolute measurements of fluorescence from different instruments. Relative values on a given measurement as a function of time, will, however, be subject to the same q.

Using Equation 13.44, we can show (see problem assignments) that

$$I_F(t) = q\Phi I_0 \left(1 - 10^{-\epsilon l c(t)}\right). \qquad (13.45)$$

If the concentration is low enough to ensure that less than about 2% of the light is absorbed, the intensity of fluorescence emission can be approximated by

$$I_F(t) = q\Phi I_0 \left(\epsilon l c(t)\right) \qquad (13.46)$$

and the fluorescence emission is therefore proportional to the concentration.

Since the emission of light can be measured against a low background (for example by observing at an angle perpendicular to the direction of excitation) it can be very sensitive. Also, a single molecule can absorb and emit light many times during a small time period providing more photons to be detected. As a consequence, fluorescence measurements can be very sensitive measures of concentration and in many cases it is possible to detect single molecules with good signal to noise.

13.3 PHOTON CORRELATION SPECTROSCOPY

Photon correlation spectroscopy (PHS), dynamic light scattering (DLS), or quasi-elastic light scattering (QELS) are different names for a technique based on fluctuations in the amount of scattering that arises from a volume illuminated by a constant light source. The scattering at any particular angle depends on the distribution of scattering elements within the volume, where the scattering can arise from differences in density across the volume or from particles, such as nanoparticles, within the volume. When the densities fluctuate or when the particles move, the scattering will change, giving rise to dynamic fluctuations in the scattering observed.

If the particles move relative to each other, the scattering that arises among these particles will interfere constructively or destructively, depending on their relative movement and orientation. The variation of the electric field of the scattered light with time is designated as $E(t)$ and the time correlation function of the field fluctuations will reflect the relative motions of the particles. It is possible to show that if the particles in solution are governed by random Brownian motion — diffusion in three dimensions — then the decay of the electric filed correlation function is exponential

$$G_E(\tau) = \langle E(t)E(t+\tau)\rangle = e^{-\Gamma\tau}. \qquad (13.47)$$

The decay constant, Γ, is in turn related to the diffusion coefficient, D, and the wave vector, q, as

$$\Gamma = Dq^2 = D\left(\frac{4\pi n}{\lambda} sin\left(\frac{\theta}{2}\right)\right)^2, \qquad (13.48)$$

where n is the refractive index, λ is the wavelength, and θ is the angle at which the scattering is observed.

Since the diffusion coefficient is inversely proportional to the radius of the particle, the decay time (the inverse of Γ) is proportional to the radius. Thus larger particles will lead to slow decays, whereas smaller particles will lead to rapid decays.

Experimentally, it is not possible to determine the fluctuations in the electric fields, but it is possible to measure the fluctuations in the intensity of the scattered light, I(t). Since the intensity is proportional to the square of the electric field, the correlation function of the intensity fluctuations can be related to the correlation function of the electric field fluctuations as

$$G_I(\tau) = G(\infty)\left[1 + \alpha\,|G_E(\tau)|^2\right] = G(\infty)\left[1 + \alpha e^{-2\Gamma\tau}\right], \qquad (13.49)$$

where $G(\infty)$ and α are constants related to the experimental baseline of the

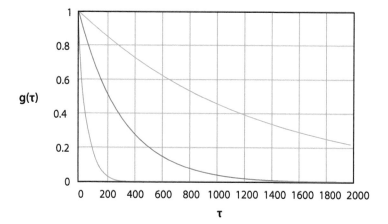

Figure 13.8: The decay of the intensity autocorrelation functions for small and large particles.

correlation function (the value as the correlation time goes to infinity) and instrumental factors that relate the electric field to the actual measured intensity. Clearly, the intensity correlation function is an equally good determinant of the particle size as illustrated in Figure 13.8.

> If you were to shine a laser pointer through a scattering solution, you would observe a pattern of scattered light on a piece of paper next to the solution. This speckle pattern will change as the particles in the solution move. The dynamic light scattering experiment is a measure of the intensity of one of the speckles as a function of time.

While the intensity of scattering depends on the concentration, the fluctuations in intensity arises primarily from the change in scattering as the relative position of the scattering elements changes rather than changes in the number of scattering elements. Thus the amplitude of the correlation functions cannot provide a direct measure of the number of scattering elements and hence Equation 13.40 is not applicable.

13.4   FLUORESCENCE CORRELATION SPECTROSCOPY

Fluorescence correlation spectroscopy (FCS) is analogous to photon correlation spectroscopy but is based on measurements of fluctuations in fluorescence emission[5] from the entities of interest. In these measurements, the fluctuations can arise from changes in emission because of fluctuations in a chemical equilibrium as well as by fluctuations arising form movement of fluorescent species in and out of the volume of observation and it is therefore a slightly more versatile technique than dynamic light scattering. However, to be most effective, the observation volume must be small, which introduces additional challenges, as will be explained in more detail in Chapter 16.

The fluorescence autocorrelation function is

$$G_F(\tau) = \langle F(t)F(t+\tau)\rangle. \tag{13.50}$$

Since the fluorescence intensity, $F(t)$, can be proportional to the concentration of emitting species as shown in Equation 13.46, the fluorescence autocorrelation function can be a direct measure of the concentration and Equation 13.40 will be applicable as will be shown in more detail in Section 16.2.

The decay function will depend on which process is the dominant process. If the intensity fluctuations arise from changes in concentration of a species because of shifts in an equilibrium, then the decay will be exponential. If, on the other hand, the intensity fluctuations arise from changes in local concentration from motion in and out of the observation volume, then the decay will depend on the nature of the motion and on the detailed shape of the observation volume, as will be shown in detail in Section 16.1.

---

[5]The details of absorption and emission of light, including the phenomenon of fluorescence is discussed in Chapters 14 and 15.

V

Understanding light-matter interactions

Introduction to Part V

A growing number of applications of nanoscale materials involve interactions of light with the material to lead to applications such as opto-electronics, sensing, diagnosis, and therapeutics. This section therefore lays the foundation for understanding how light and matter interact, how absorption and emission are related and how that can provide insights. This section therefore relates back to the quantum confinement phenomena described in Part II.

Chapter 14 reviews the first order quantum mechanical description of the processes of absorption and emission, setting the foundation for understanding the processes of absorption and emission in molecular systems which is the subject of Chapter 15. This chapter focuses in particular on the principles of fluorescence since this is an increasingly useful tool for studying properties of materials and it allows for understanding energy transfer processes. Chapter 16 introduces the tool of fluorescence microscopy to study fluctuations to study dynamics as discussed in Chapter 13. Finally, Chapter 17 summarizes some of the principles behind the interaction of light with matter as a general background for understanding properties such as surface plasmon polaritons.

CHAPTER 14

The principles of absorption and emission of light

A number of tools used to explore the world of nanoscience and nanotechnology rely on probing the properties of matter with electromagnetic radiation, or light, at various wavelengths. The use of absorption or emission of light is generically known as spectroscopy. Depending on the wavelength, or the energy of the radiation, the light will probe different properties of the molecules or the material of interest. Table 14.1 provides an overview of the relationship between the wavelength, frequency, and the energy of the electromagnetic radiation and the corresponding energy transitions.

The electromagnetic radiation spans fifteen orders of magnitude in wavelength (picometer to kilometer), frequency (kilohertz to exahertz), and energy (femtoelectron volts to gigaelectron volts) corresponding to energy transitions ranging from nuclear transitions, to electronic transitions, molecular

Table 14.1: Electromagnetic Spectrum in Wavelength, Frequency, and Energy Units (the numbers indicate upper or lower limits).

Name	Wavelength	Frequency	Energy	Transitions
γ rays	$< 10 \ pm$	$> 30 \ EHz$	$> 100 \ keV$	nuclear
X-rays	$\rightarrow 10 \ nm$	$\leftarrow 30 \ PHz$	$\leftarrow 0.12 \ keV$	inner electron
UV	$\rightarrow 390 \ nm$	$\leftarrow 0.8 \ PHz$	$\leftarrow 3.3 \ eV$	valence electron
Visible	$\rightarrow 750 \ nm$	$\leftarrow 400 \ THz$	$\leftarrow 1.7 \ eV$	valence electron
IR	$\rightarrow 1 \ mm$	$\leftarrow 0.3 \ THz$	$\leftarrow 1.2 \ meV$	vibrational
microwave	$\rightarrow 10 \ mm$	$\leftarrow 0.3 \ GHz$	$\leftarrow 0.12 \ meV$	rotational
UHF	$\rightarrow 100 \ mm$	$\leftarrow 3 \ GHz$	$\leftarrow 12 \ \mu eV$	electron spin
VHF	$\rightarrow 1 \ m$	$\leftarrow 0.3 \ GHz$	$\leftarrow 1.2 \ \mu eV$	nuclear spin

motions, and spin transitions. The corresponding spectroscopic tools can therefore probe just about every material property we may be interested in. While
the energies and the material properties being probed vary greatly, the underlying principles of absorption or emission of electromagnetic radiation is the
same for all. The purpose of this chapter is to describe some of the simplest
of these principles.

14.1 THE QUANTUM MECHANICAL DESCRIPTION OF AB-
SORPTION AND EMISSION OF ENERGY

Consider a system that is described by a set of wave functions (eigenfunctions
of the Hamiltonian in the time-dependent Schrödinger Equation):

$$\psi_j(q,t) = \phi_j(q)e^{-iE_j t/\hbar}. \tag{14.1}$$

Here, $\phi_j(q)$ is the stationary state wave functions (solutions to the time-
independent Schrödinger Equation).

Consider two states given by the quantum numbers m and n respectively,
with $E_m < E_n$. Since the individual wave functions are solutions to the
Schrödinger Equations, any linear combination will be as well. Thus, the wave
function

$$\psi_{mn}(q,t) = c_m(t)\psi_m(q,t) + c_n(t)\psi_n(q,t) \tag{14.2}$$

is also an eigenfunction of the time-dependent Hamiltonian. In this linear
combination, the coefficients $c_m(t)$ and $c_n(t)$ represent the proportion of each
of the two states that contribute to the combined wave function as a function
of time. They will represent the probability of one or the other state being
occupied at a particular time, t, when they are subjected to a perturbation
that can couple the two states.

Consider the system described by a time-independent Hamiltonian, $\hat{H}_s$,
being subjected to an external perturbation described by a new part of the
Hamiltonian, $\hat{H}_{pert}(t)$, corresponding to a time-varying external electric field,
$E(t)$ as present in the electromagnetic radiation. This external perturbation
can couple the two states and provide the energy needed to transition from
one to the other. We will now have to solve the time-dependent Schrödinger
Equation for the system including the perturbation component, i.e., we need
to solve

$$\left[\hat{H}_s + \hat{H}_{pert}(t)\right]\psi_{mn}(q,t) = i\hbar\frac{\partial}{\partial t}\psi_{mn}(q,t). \tag{14.3}$$

Substituting Equation 14.2 leads to

$$\left[\hat{H}_s + \hat{H}_{pert}(t)\right] \left[c_m(t)\psi_m(q,t) + c_n(t)\psi_n(q,t)\right]$$
$$= i\hbar \frac{\partial}{\partial t} \left[c_m(t)\psi_m(q,t) + c_n(t)\psi_n(q,t)\right]. \qquad (14.4)$$

Following the differentiation of each term on the right hand side, we get

$$\left[\hat{H}_s + \hat{H}_{pert}(t)\right] \left[c_m(t)\psi_m(q,t) + c_n(t)\psi_n(q,t)\right]$$
$$= i\hbar \left[c_m(t)\frac{\partial}{\partial t}\psi_m(q,t) + c_n(t)\frac{\partial}{\partial t}\psi_n(q,t)\right]$$
$$+ i\hbar \left[\psi_m(q,t)\frac{\partial}{\partial t}c_m(t) + \psi_n(q,t)\frac{\partial}{\partial t}c_n(t)\right]. \qquad (14.5)$$

The first term on the right hand side is the solution to the eigenvalue problem of $\hat{H}_s\psi_{mn}(q,t)$, so the second term must reflect the effect of the perturbation Hamiltonian, so that

$$\left[\hat{H}_{pert}(t)\right] \left[c_m(t)\psi_m(q,t) + c_n(t)\psi_n(q,t)\right]$$
$$= i\hbar \left[\psi_m(q,t)\frac{\partial}{\partial t}c_m(t) + \psi_n(q,t)\frac{\partial}{\partial t}c_n(t)\right]. \qquad (14.6)$$

In order to progress, we use the trick of first multiplying both sides of Equation 14.6 by the complex conjugate of the wave function corresponding to state n, i.e., $\psi_n^*(q,t)$ and then integrating over all space to get

$$c_m(t)\int \psi_n^*(q,t)\hat{H}_{pert}(t)\psi_m(q,t)dq + c_n(t)\int \psi_n^*(q,t)\hat{H}_{pert}(t)\psi_n(q,t)dq$$
$$= i\hbar \int \psi_n^*(q,t)\psi_m(q,t)\frac{\partial}{\partial t}c_m(t)dq + i\hbar \int \psi_n^*(q,t)\psi_n(q,t)\frac{\partial}{\partial t}c_n(t)dq. \qquad (14.7)$$

We know that the wave functions are orthonormal, which means that the wave functions are orthogonal and normalized, such that

$$\int \psi_n^*(q,t)\psi_m(q,t)dq = 0 \qquad (14.8)$$
$$\int \psi_n^*(q,t)\psi_n(q,t)dq = 1. \qquad (14.9)$$

Accordingly, the first term of the right hand side of Equation 14.7 is zero and the second term reduces to the partial derivative of $c_n(t)$ with respect to time.

If we assume that when the perturbation is initiated at $t = 0$, the system is exclusively in the mth state, then $c_n(0) = 0$ and $c_m(0) = 1$, which in turn means that the second term on the left hand side of Equation 14.7 is zero, and this equation then reduces to

$$\int \psi_n^*(q,t)\hat{H}_{pert}(t)\psi_m(q,t)dq = i\hbar\frac{\partial}{\partial t}c_n(t) \tag{14.10}$$

Thus if we can calculate the effect of the perturbation Hamiltonian on the wave function of the initial state, we can determine the rate of change of the population of the final state. Equation 14.10 is therefore a general relation for determining the change in the fraction of two states that are coupled by a perturbation.

14.1.1   The transition dipole moment for electronic transitions

The electric field of the electromagnetic radiation varies as a function of time as a sinusoidal function with frequency ν generally described as

$$\vec{E}(t) = \vec{E}_0\left(e^{ih\nu t/\hbar} + e^{-ih\nu t/\hbar}\right). \tag{14.11}$$

Note that $\omega t = 2\pi\nu t = h\nu t/\hbar$, where $h\nu$ represents the energy associated with the electromagnetic ration.

This oscillating electric field will, in the case of electronic transitions, interact with the electrons in the systems whose net distribution is characterized by a dipole moment, $\vec{\mu}$.[1] The dipole moment has a magnitude proportional to the net charge separation and the distance of separation and is generally expressed as $\vec{\mu} = e\vec{r}$ where $\vec{r}$ determines the separation of the charges and the orientation of the dipole moment in a coordinate system tied to the system.

The interaction between the oscillating electric field and the electrons in the system is classically expressed by a Hamiltonian that is the dot product of the electric field vector and the dipole moment vector, $H = \vec{E} \cdot \vec{\mu}$ and the corresponding perturbation Hamiltonian in Equation 14.10 is

$$\hat{H}_{pert} = \hat{E} \cdot \hat{\mu} = E_0\left(e^{ih\nu t/\hbar} + e^{-ih\nu t/\hbar}\right)\hat{\mu}. \tag{14.12}$$

Since the wave functions are products of a spatial and time dependent component as shown in Equation 14.1 it is easy to show that Equation 14.10 reduces to

---

[1]In the case of electron or nuclear spin transitions as in electron paramagnetic resonance or nuclear magnetic resonance, it is the magnetic field of the electromagnetic radiation that interacts with the electron or nuclear magnetic moments.

$$E_0 \left(e^{ih\nu t/\hbar} + e^{-ih\nu t/\hbar} \right) \cdot \left(e^{(E_n - E_m)t/\hbar} \right) \int \phi_n^*(q)\hat{\mu}\phi_m(q)dq$$

$$= i\hbar \frac{\partial}{\partial t} c_n(t). \tag{14.13}$$

The integral in Equation 14.13 is called the **transition dipole moment** for the transition from state m to state n and is given the symbol μ_{nm}

$$\mu_{nm} = \int \phi_n^*(q)\hat{\mu}\phi_m(q)dq. \tag{14.14}$$

It is significant to note that the transition dipole moment is defined in terms of the stationary state wave functions, which is therefore independent of time. It is therefore possible to integrate Equation 14.13 over time to get an expression for the variation in $c_n(t)$, which is related to the fraction of the nth state that is populated as a function of time after the onset of the perturbation at $t = 0$. The integration yields

$$c_n(t) = E_0 \left[\frac{1 - e^{i(E_n - E_m + h\nu)t/\hbar}}{E_n - E_m + h\nu} + \frac{1 - e^{i(E_n - E_m - h\nu)t/\hbar}}{E_n - E_m - h\nu} \right] \mu_{mn}$$

$$\approx E_0 \left[\frac{1 - e^{i(E_n - E_m - h\nu)t/\hbar}}{E_n - E_m - h\nu} \right] \mu_{mn}. \tag{14.15}$$

The approximation in Equation 14.15 arises whenever the energy of the electric field of the electromagnetic radiation $h\nu$ is close to the energy difference between the two states, i.e., whenever the electromagnetic radiation is on resonance. Close to resonance, the denominator $E_n - E_m - h\nu \to 0$, so the second term will dominate.

The probability of the system being in the nth state at any given time is related to the square of Equation 14.15

$$|c_n^2| = c_n^*(t)c_n(t) = 4\pi E_0^2 \mu_{mn}^2 \frac{sin^2\left(\frac{1}{2}(E_n - E_m - h\nu)\right)}{(E_n - E_m - h\nu)^2}. \tag{14.16}$$

14.1.2 Selection rules for electronic transitions

The transition dipole moment as defined by Equation 14.14 is an integral over space. It will therefore not exist, be equal to zero, if the total integrand is antisymmetric with respect to the spatial coordinates. The relative symmetry of the two states and the dipole moment is therefore important. If the product $\phi_n^*(q)\hat{\mu}\phi_m(q)$ is symmetric, the transition from the mth state to the nth state is an allowed transition. The extent to which the symmetry is present will then determine the extent to which the transition is allowed.

The concept of forbidden and allowed transitions is illustrated most simply if we assume that the dipole operator is arranged along the x-axis such that $\hat{\mu} = e\hat{x} = ex$ since the problem reduces to a one-dimensional problem. In this case, the operator is odd, with respect to the x-coordinate. Thus, if the wave functions are both even functions, then the total integrand is odd and the integral is zero and the transition is a forbidden transition. Conversely, if one wave function is even and the other is odd, then the total integrand is even and the transition is allowed.

A function is even if it has the same value for positive and negative coordinates and it is odd if the value has the opposite signs for positive and negative coordinates

$$Even\ function: f(x) = \ \ f(-x)$$
$$Odd\ function: f(x) = -f(-x).$$

In general then, when $\phi_n^*(q)$ and $\phi_m(q)$ are both even or both odd

$$\mu_{mn} = e \int \phi_n^*(q)\hat{\mu}\phi_m(q)dq = 0 \tag{14.17}$$

and the transition is forbidden, while when $\phi_n^*(q)$ and $\phi_m(q)$ have opposite symmetry

$$\mu_{mn} = e \int \phi_n^*(q)\hat{\mu}\phi_m(q)dq \neq 0 \tag{14.18}$$

and the transition is allowed.

Consider as a specific example the wave functions for the simple harmonic oscillator (Section 4.3.3.2) given by Equation 4.84 and illustrated in Figure 14.1 for the quantum numbers $n = 0, 1, 2, 3,$ and 4. It is evident that the wave functions corresponding to $n = 0, 2,$ and 4 are even, while those for $n = 1$ and 3 are odd. Thus the transitions are allowed whenever m and n differ by an odd number: $\mu_{mn} \neq 0$ when $(m - n = 1, 3, \ldots)$, but are forbidden whenever they differ by an even number: $\mu_{mn} = 0$ when $(m - n = 2, 4, \ldots)$

14.1.3 The Einstein coefficients

The transition dipole moment as described in Equation 14.14 refers to the transition from the mth state to the nth state, but does not explicitly describe which of these states is the higher energy state and which is the lower energy state. It is therefore also possible to define a corresponding transition dipole moment for the reverse transition, i.e.,

$$\mu_{mn} = \int \phi_m^*(q)\hat{\mu}\phi_n(q)dq. \tag{14.19}$$

$\varphi_n(x)$

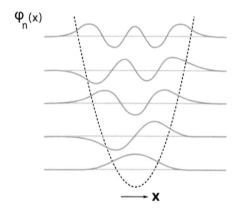

Figure 14.1: The first five wave functions for the simple harmonic oscillator.

If we have a linear dipole moment operator, as in the example above for the electric field of the electromagnetic radiation, then we expect $\mu_{mn} = \mu_{nm}$. If a transition is stimulated by an applied electromagnetic field and takes the system from a lower energy state to a higher energy state, the transition is termed **stimulated absorption**,[2] while if a transition is stimulated by an applied electromagnetic field and takes the system from a higher energy state to a lower energy state, the transition is termed **stimulated emission**.[3] In the first case the system absorbs a photon, while in the second case the system emits a photon in addition to the photon that caused the stimulated emission.

In most cases the system of interest is imbedded in a surrounding medium in which there are molecules that constantly move (by translation, vibration, or rotation). There will therefore always a background of randomly oscillating electric (or magnetic) fields present around the system (see the concept of black body radiation later). It is therefore possible that the system will spontaneously absorb energy from the surroundings to cause a transition from a lower energy state to a higher energy state or spontaneously emit energy to the surroundings to cause a transition from a higher energy state to a lower energy state. This will lead to an equilibrium distribution given by the Boltzmann distribution (for non-degenerate states)

$$\frac{N_n}{N_m} = e^{-(E_n - E_m)/kT} = e^{-\Delta E/kT}. \tag{14.20}$$

In the case of electronic transitions, the energy difference is so large that the fraction in Equation 14.20 is nearly zero and there are no systems in the higher energy states in the absence of applied external fields. However,

---

[2]alternatively induced absorption or just absorption.
[3]alternatively induced emission.

for lower energy transitions, such as vibrations or rotations, there can be a significant population in the higher energy states. In fact, for the lowest energy transitions associated with spin transitions, which frequently have only two states, the population distribution is nearly the same in the two states.

> Recall that $kT \approx 4 \times 10^{-23} J$ which corresponds to about 26 meV. This value may be compared to energies associated with the various transitions listed in Table 14.1. The ambient energy per degree of freedom is, according to the equipartition theorem, about $1/2kT$, which is comparable to some of the lowest vibrational transitions, but also much less than many vibrational transitions.

Application of an external electromagnetic radiation perturbs the equilibrium distribution, but with continuous application, will lead to a new equilibrium where there is a balance between the absorption on the one hand and the stimulated and spontaneous emissions on the other hand.

Using the standard convention, we will define the Einstein coefficients as follows:

B_{mn} as the transition probability per unit time for stimulated absorption,
B_{nm} as the transition probability per unit time for stimulated emission,
A_{mn} as the transition probability per unit time for spontaneous absorption,
A_{nm} as the transition probability per unit time for spontaneous emission.

In many cases, and certainly for electronic transitions, the spontaneous absorption has a very small transition probability so for all practical purposes, $A_{mn} \approx 0$. Likewise, the transition probability for stimulated absorption and emission is the same (or very similar) since the transition dipole moments are the same, so $B_{mn} = B_{nm}$ (for non-degenerate states).

Thus the rate of change in the number of systems in the nth state as a result of stimulated absorption, stimulated emission, and spontaneous emission, respectively, will be

$$\left(\frac{dN_n}{dt}\right)_{stim\ abs} = N_m B_{mn}\rho(\nu) \qquad (14.21)$$

$$\left(\frac{dN_n}{dt}\right)_{stim\ em} = -N_n B_{nm}\rho(\nu) \qquad (14.22)$$

$$\left(\frac{dN_n}{dt}\right)_{spon\ em} = -N_n A_{nm}\rho(\nu). \qquad (14.23)$$

where $\rho(\nu)$ is a function that describes the density of radiation as a function of frequency.

When the new equilibrium has been established, the net rate of absorption will be equal to the net rate of emission, so that the sum of these rates adds to zero

$$\left(\frac{dN_n}{dt}\right)_{net} = N_m B_{mn}\rho(\nu) - N_n B_{nm}\rho(\nu) - N_n A_{nm} = 0. \qquad (14.24)$$

Solving for A_{nm} gives

$$A_{nm} = \frac{N_m - N_n}{N_n}\rho(\nu)B_{nm} = \left(e^{\Delta E/kT} - 1\right)\rho(\nu)B_{nm}. \qquad (14.25)$$

The radiation density will be defined by the illumination source. In the case of black body radiation (see later), the radiation density is given by

$$\rho(\nu) = \frac{8\pi h\nu^3}{c^3}\frac{1}{e^{h\nu/kT} - 1}, \qquad (14.26)$$

where c is the speed of light.

This then provides for a simple, and general, relationship between the two Einstein coefficients

$$A_{nm} = \frac{8\pi h\nu^3}{c^3}B_{nm}. \qquad (14.27)$$

The end result is that if one of the three Einstein coefficients are known, then all three are known (if the system has degeneracies of g_m and g_n, the relationship between the B-coefficients is more generally $g_n B_{nm} = g_m B_{mn}$). Moreover, since whenever the only path of transition from the nth state to the mth state is by spontaneous emission, the decay will be exponential with a spontaneous radiative lifetime, $t_{sponem} = 1/A_{nm}$, which can be measured. There is therefore an experimental path towards understanding the rates of stimulated absorption and emission processes for particular radiation fields as given by Equations 14.20-14.23.

14.1.4 Absorption and emission line shapes

While we have described the transitions in terms of defined energy levels, E_m and E_n, we recognize that the there is a range of energies of electromagnetic radiation that can lead to stimulated absorption or emission. Moreover, the expression in Equation 14.16 shows that the probability of transition is finite even when $h\nu \neq (E_n - E_m)$. This is illustrated on Figure 14.2 for the function $sin^2 x/x^2$. This is therefore the expected line shape for the absorption process to a first approximation.

Consider a case where a the nth state is populated to the level of N_n^0 by a short pulse of exposure to electromagnetic radiation before the system can reach equilibrium. After the pulse, the system can only return to the original state by spontaneous emission through Equation 14.23, which leads to a exponential decay with a time constant of $1/A_{nm}$ or frequency A_{nm}

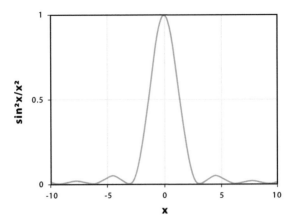

Figure 14.2: Illustration of the line shape for sin^2x/x^2.

$$N_n(t) = N_n^0 e^{-A_{nm}t}. \tag{14.28}$$

The frequency dependence of the emission process (and by symmetry the absorption process) can be obtained by finding the Fourier transform of the decay function, i.e.,

$$g(\nu) = \frac{1}{\sqrt{2\pi}} \int_0^\infty e^{-A_{nm}t} e^{-2\pi i\nu t} dt = \frac{1}{A_{nm} - 2\pi i\nu}. \tag{14.29}$$

This is a complex function with a real and an imaginary part. The real part is called the Lorentzian line shape function, and the normalized function is given by

$$g(\nu) = \frac{1}{\pi} \frac{A_{nm}}{A_{nm}^2 + (2\pi\nu)^2}. \tag{14.30}$$

In the absence of any other effects, the frequency dependence of the absorption or emission spectra will follow the Lorentzian line shape as illustrated in Figure 14.3. The half-width at half-height is given by A_{nm} and is expected to exceed that of the line shape function in Figure 14.2.

14.1.4.1 Homogeneous and heterogeneous line broadening

The Lorentzian line shape arises from the transition between two well-defined energy levels. If all the molecules or entities in the sample are identical, the observed line shape is termed the **homogeneous line shape** because it corresponds to a homogeneous population. In general, however, we expect the energy levels to contain some uncertainties and hence there will be a range of

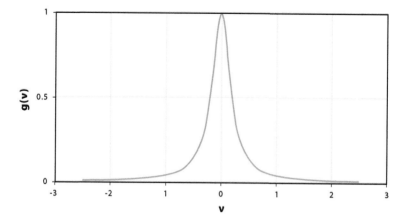

Figure 14.3: Illustration of the Lorentzian line shape.

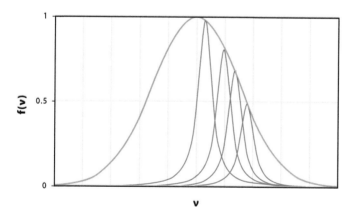

Figure 14.4: Illustration of the heterogeneous line shape.

absorptions. Likewise, the detailed environment around molecules or nanostructures will vary across the sample also giving rise to a range of absorptions. The net effect is that the observed line shape will be a sum of a series of Lorentzian lines as illustrated in Figure 14.4. This is now termed a **heterogeneous line shape**.

In general we expect the overall line shape to be a convolution of the Lorentzian line shape with a normal or Gaussian distribution, which is called a Voigt line shape. Whenever the Lorentzian line width is much smaller than the Gaussian distribution, the line shape approaches a Gaussian line shape.

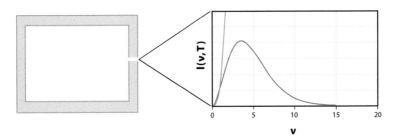

Figure 14.5: Illustration of the origin of black body radiation and the spectrum (red curve) as a function of the frequency (in units of 10^{14} Hz).

14.1.5 Black-body radiation

Consider a box, as illustrated in Figure 14.5 (the shape is not critical), with a tiny whole through which one can measure the intensity of electromagnetic radiation as a function of the energy (as measured by either the frequency or the wavelength). Experiments of this nature showed that the distribution of energy followed a curve where the intensity approached zero at both low and high frequency and went through a maximum, whose position depended on the temperature of the box, as seen in the red curve in Figure 14.5.

The explanation for this observation was that the movement of electrons in the material created electromagnetic radiation that was emitted at all frequencies. The interpretation based on classical physics, where all frequencies of oscillation of electrons were possible, demanded a dependence on the frequency cubed, as illustrated in the blue curve in Figure 14.5 and could not explain the decay at higher frequencies. In order to explain the observation, Planck had to postulate that the frequencies of oscillations were discrete rather than continuous and that the discrete levels were proportional to a constant, h, later called Planck's constant. In order to fit the data, he postulated that the intensity distribution would be

$$I(\nu, T) = \frac{2h\nu^3}{c^2} \frac{1}{e^{h\nu/kT} - 1}. \tag{14.31}$$

To fit the data, at various temperatures, he needed $h = 6.6 \times 10^{-34} J$ s. The explanation of the black body radiation was therefore the dawn of quantum mechanics.

CHAPTER 15

Molecular spectroscopy

The previous chapter focused on the quantum mechanical description of the interaction of light with matter with particular attention to electronic transitions. In this chapter, we will explore some of the consequences and utilities of electronic transitions in probing properties of molecules through various forms of spectroscopy and then show how these can be applied to study nanoscale phenomena. The emphasis will be on the concept of fluorescence spectroscopy since this has proven particularly useful in studies in the life sciences with applications in nanobiotechnology and bionanotechnology.

> Some distinguish the concept of **nanobiotechnology** as the application of nanoscale tools to understand biological systems from the concept of **bionanotechnology** as the application of biological understanding to nanotechnology systems. An example of the former is the use of electron microscopes to study biological structures and an example of the latter is the developing of self-assembly of nanostructures based on our understanding the biological macromolecule assembly.

15.1 THE OPTICAL TRANSITIONS

The electronic transitions in many molecules and in some nanomaterials occur in the optical part of the electromagnetic spectrum. Consider the case of conjugated electronic systems described in section 5.1.1, where the energy needed to excite an electron from the ground electronic state to the first excited electronic state corresponded to wavelengths ranging from 200 to 300 nm, that is, in the ultra-violet part of the electromagnetic spectrum. In many organic molecules the conjugations extend the "particle-in-the-box" so that the energy separations decrease and the wavelengths extend to the 400-700 nm range, which is the visible part of the electromagnetic spectrum.

Since the electrons are associated with bonds within the molecular structure, the energies are affected by the vibrational state of the bond so that

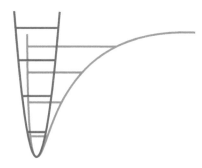

Figure 15.1: Schematic comparison of the harmonic (red) and anharmonic (blue) potentials.

illumination of the molecule will affect both the electronic state and the vibrational state of the molecule. Recall that vibrational energy separations, as described for example by the simple harmonic oscillator, are much smaller than electronic transitions, and are usually in the infra-red part of the electromagnetic spectrum corresponding to wavelengths greater than 1000 nm.

While the simple harmonic oscillator is a good first approximation to the vibration of a bond, it fails to consider that the bond vibrations are necessarily anharmonic in the sense that as the bond lengths are increased in higher vibrational states, there will be energies above which the bond will break and the molecule dissociates into two parts. This anharmonic potential is compared schematically with the harmonic potential in Figure 15.1.

There are several important differences. First, the potential goes to zero as the bond length increases; second, the potential goes to infinity as the bond length decreases; third, the number of vibrational states is finite; fourth, the minimum in the potential is expected to increase to large intermolecular distances since the electron most frequently is moved from a bonding orbital to a non-bonding or anti-bonding orbital with a lower bond order; and fifth, the difference in energy between the vibrational states decreases as the quantum number increases, in contrast to the harmonic potential where the difference in energy between neighboring states is independent of the quantum number (see Section 4.3.3.2)

The anharmonic potential will also exist in the electronically excited states as illustrated in Figure 15.2.

The diagrams in Figure 15.2 suggest that electromagnetic radiation at the appropriate energy can lead to a simultaneous change in the electronic

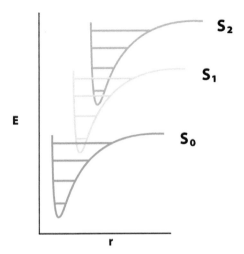

Figure 15.2: Schematic comparison of the anharmonic potentials in the ground state (S_0 dark blue), the first electronic excited state (S_1 light blue) and the second electronic excited state (S_2 light gray).

and vibrational state of the molecule as long as the selection rules make the transition allowed (Section 14.1.2).

An additional refinement of this picture is that at each of the vibrational levels, there are a set of rotational states which are at even smaller energy difference and therefore in effect cause a broadening of the vibrational states.

15.1.1 The Frank–Condon Principle

The diagram in Figure 15.3 shows the transition that can occur from the lowest electronic (S_0) and vibrational state and ($v = 0$) to the first electronic state (S_1) and first vibrational state ($v' = 1$). Since we expect the interaction between the electromagnetic radiation and the electronic state to be very rapid, the electronic transition is expected to be much faster (on the order of femtoseconds) than the actual movement of the nuclei in the molecule (on the order of picoseconds). This leads to the **Frank–Condon Principle**, which states that the electronic transition occurs without changes in the value of the nuclear coordinates along the horizontal nuclear coordinate axis. Accordingly, we refer to this as a **vertical transition** as indicated by the green arrow in Figure 15.3.

Following the electronic transition, the nuclei can move and the molecule can relax to the lowest available vibrational state within the electronically excited potential. This process is referred to as **internal conversion** and involves both vibrational and rotational movements on the picosecond time scale.

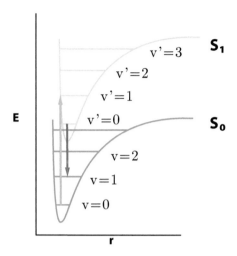

Figure 15.3: Illustration of the Frank–Condon Principle for electronic transitions.

At some time later, the electronically exited state can return to the electronic ground state, either by stimulated or spontaneous emission or by other relaxation processes. The emission process must also follow the Frank–Condon Principle and result in a vertical transition as illustrated by red arrow in Figure 15.3, in this case from the vibrational ground state ($v' = 0$) in the electronically excited state (S_1) to the excited vibrational state ($v = 1$) in the ground electronic state (S_0). This is then followed by internal conversion in the ground state towards the lower vibrational state.

15.1.2 Intersystem crossing

In most cases, there are two electrons in the ground electronic states with opposite spins, this is called the **singlet** state, hence the use of the symbol S_0. When the electron is excited to a higher, unoccupied, electronic state, the selection rules demand that the spin does not change. The allowed electronic transitions induced by electromagnetic radiation is therefore to excited singlet states, such as S_1 and S_2.

While the Pauli Principle demands that two spins in the same state be of opposite spins, electrons in different states prefer to be of the same spin, forming a **triplet** state, T_1 which will be of lower energy than the corresponding singlet state, as illustrated schematically in Figure 15.4.

Although the singlet-triplet transition is spin forbidden, it can be coupled with other spin transitions in the surroundings, so there is a finite probability in some molecules that the excited singlet state converts spontaneously to a triplet state. This process is called **intersystem crossing** and is illustrated

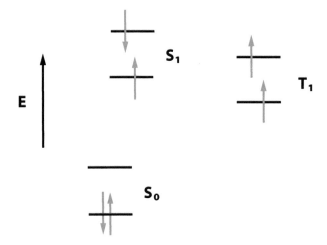

Figure 15.4: Illustration of the relationship between the ground singlet state, S_0, the first excited singlet state, S_1, and the lowest energy triplet state T_0

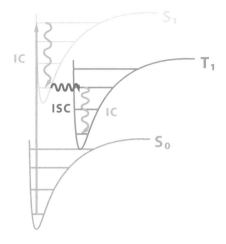

Figure 15.5: Illustration of the relationship between the ground singlet state, S_0, the first excited singlet state, S_1, and the lowest energy triplet state T_0.

in Figure 15.5. Note that the intersystem crossing is isoenergetic and would be to an excited vibrational state of the triplet state.

The concept illustrated schematically in Figure 15.5 is the process whereby a molecule is excited into an excited singlet electronic and an excited vibrational state (green arrow), followed by an internal conversion to the lowest

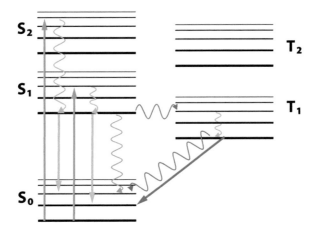

Figure 15.6: An example of the Jablonski diagram.

vibrational state of the singlet excited state (dark green wavy arrow), followed by an intersystem crossing to an excited vibrational state in the triplet
state (red wavy arrow), followed finally by an internal conversion to the lowest
vibrational state in the triplet state.

15.1.3 The Jablonski diagram

In order to simplify the description of the many processes that arise during
and after a transition from the ground state to an excited state, we frequently
choose to omit the details of the potential and just depict the various energy
states as shown in Figure 15.6 where it is understood that the positions of
the various energy levels are for illustration and are not necessarily to scale
in either the energy dimension or in the coordinate dimension.

This simplified version of the energy levels is called the Jablonski diagram

Figure 15.6 illustrates many of the processes that are important in the
excitation-relaxation cycle of optical transitions depicting the electronic and
vibrational states schematically.[1]

The blue vertical arrows depict two possible excitations, usually from the
ground electronic and vibrational state (S_0) to excited electronic and vibrational states (S_1 and S_2) according to the Frank–Condon Principle. The time
scale for these transitions is on the order of femtoseconds.

Following the excitation, there are rapid internal conversion processes, indicated by the orange wavy arrows, that cause relaxation to the lowest available

---

[1]For simplicity we do not show the possible rotational states, but it is understood that
these would be present and that they fill much of the gaps between the vibrational states.

vibrational state in the excited electronic state manifold at a time scale of picoseconds.

Once the molecule is in the lowest vibrational state of the excited electronic state, there are several possible processes available. First, there may be a non-radiative relaxation from the excited electronic state to the ground electronic state via the vibrational and rotational manifold, as indicated by the blue wavy arrow. Second, there may be a *spontaneous emission* of a photon by a process we term **fluorescence** as indicated by the green vertical arrows, which show that the emission may be to distinct vibrational states in the ground state, depending on the Frank–Condon Principle. Third, there can be a transition from the excited singlet electronic state (S_1) to the lowest triplet electronic state (T_1), as shown by the horizontal red wavelet. This **intersystem crossing** process is, in principle, forbidden and the probability that this occurs is normally much smaller than the non-radiative relaxation or fluorescence processes. The lifetime of the lowest excited electronic state is usually on the nanosecond time scale.

If the intersystem crossing occurs, it is followed by a rapid internal conversion within the triplet state. Subsequently, the triplet state can relax to the ground singlet state by another non-radiative relaxation process, which also requires an intersystem crossing (as shown by the angled red wavy arrow). Alternatively, there can be a spontaneous emission of a photon by a process we term **phosphorescence** as indicated by the angled red arrow. Since both of these processes are of lower probability, the lifetime of the triplet state (T_1) is much longer, typically microseconds and longer.

It is evident from the diagram in Figure 15.6 that the energy of the photon required for excitation must be greater than or equal to the energy of the photon emitted by fluorescence which in turn is greater or equal to the energy of the photon emitted by phosphorescence.

15.2 FLUORESCENCE SPECTROSCOPY

The Jablonski diagram in Figure 15.6 suggests that the lowest energy of excitation from the vibrational state in the ground electronic state ($S_0; v = 0$) to the lowest vibrational state of the first excited electronic state ($S_1; v' = 0$) should be equal to the highest energy of emission from ($S_1; v' = 0$) to ($S_0; v = 0$). The start of the absorption spectrum should therefore be at the same energy as the start of the emission spectrum. The next higher energy of excitation will be at an energy higher by the difference in energy between $v' = 0$ and $v' = 1$ while the next lower energy of emission will be at energy lower by the difference in energy between $v = 0$ and $v = 1$. This trend will continue as illustrated schematically in Figure 15.7. Within the approximation that the vibrational levels in the ground and excited electronic states are the same, we expect these transitions to give absorptions and emissions at energies that are mirrored about the energy of the central transition as shown diagrammatically in Figure 15.7. In reality, each of these transitions are broadened by energy un-

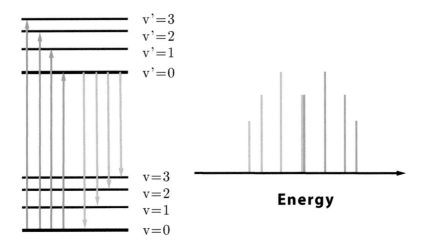

Figure 15.7: Schematic illustration of the excitation and emission processes and the corresponding expected spectrum.

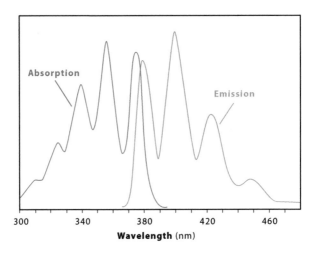

Figure 15.8: The absorption and emission spectra for anthracene.

certainties and rotational energy levels, but the principle is observed in many cases where the vibrational transitions can be resolved. This is illustrated for the absorption and emission spectra for the molecule anthracene in Figure 15.8 which is plotted as a function of wavelength rather than energy. Anthracene is a rigid molecule, which is why the vibrational levels are well resolved.

15.2.1 Absorption and emission kinetics

The field of fluorescence spectroscopy is essentially the study of the relationship between the absorption of light and the fate of the excited states that arise. We will therefore focus our attention on the process of preparing the excited state and the processes whereby the excited state can disappear.

15.2.1.1 Preparing the excited state

The absorption of light depends on the intensity of light and on the extent to which the transition is allowed. The latter is deduced from the transition dipole moment which is in turn reflected in the experimental quantity called the absorption coefficient, k_ν. This can be described in terms of the Einstein coefficients (B_{nm} and B_{mn}) and the radiation field as shown in Equation 15.1.

$$\begin{aligned}
k_\nu &= \frac{h\nu_{nm}}{4\pi}\left(N_n B_{nm} - N_m B_{mn}\right)\rho(\nu) \\
&= \frac{h\nu_{nm}}{4\pi}N_n B_{nm}\left(1 - \frac{g_n N_m}{g_m N_n}\right)\rho(\nu) \\
&= \frac{h\nu_{nm}}{4\pi}N_n B_{nm}\left(1 - e^{-h\nu/kT}\right)\rho(\nu)
\end{aligned} \tag{15.1}$$

The absorption coefficient reflects the number of photons absorbed per unit length of the material, so the integration of the absorption coefficient along the path of the light will yield the optical density, τ_ν,

$$\tau_\nu = \int k_\nu(z)dz. \tag{15.2}$$

If the incident intensity of light at some frequency, ν, at the surface of the material is $I_\nu(0)$, then we expect the the intensity at some depth x in the material is attenuated in an exponential fashion to

$$I_\nu(x) = I_\nu(0)e^{-\tau_\nu} \tag{15.3}$$

or

$$\tau_\nu = -ln\frac{I_\nu(x)}{I_\nu(0)}. \tag{15.4}$$

Accordingly, τ_ν is called the attenuation coefficient.

In the case of solutions, we expect the attenuation to be dependent on the concentration of the material in solution as well as the thickness of the sample, l. In that case, we describe the attenuation coefficient in terms of a molar extinction coefficient, ϵ, the molar concentration, C, and the thickness, l, and call it the optical density, OD, of the sample. Thus,

$$\tau_\nu = OD = -ln\frac{I_\nu(l)}{I_\nu(0)} = \epsilon l C. \tag{15.5}$$

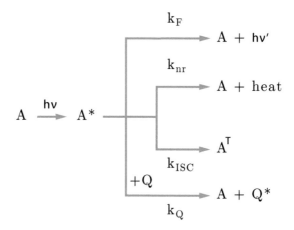

Figure 15.9: Some possible decay paths for a molecule in an excited state.

The molar extinction coefficient has units of inverse length and inverse molarity, i.e., $m^{-1}M^{-1}$ or $m^2 mol^{-1}$. While it is in principle possible to calculate the absorption coefficient, most are experimentally determined.

15.2.1.2 Decay of the excited state

Once formed, the excited state can disappear through first order decay processes such as non-radiative relaxation, intersystem crossing or radiative emission (fluorescence). In addition, the excited state may interact with other molecules in the neighborhood to facilitate a non-radiative relaxation process. This we call **quenching**.

Figure 15.9 illustrates some of the parallel process that can cause an excited state, here denoted A^*, to decay to the ground state A or the triplet state A^T.

The rate equation for each of the first order decay processes are the same

$$\frac{dA^*}{dt} = -k_x[A^*].\tag{15.6}$$

This corresponds for each of these processes to an exponential decay of the excited state and hence

$$[A^*] = [A_0]e^{-k_x t}.\tag{15.7}$$

If the concentration of the quencher, Q, is approximately constant, it also will result in a exponential decay of the excited state.

Since these are parallel processes, the overall rate of disappearance of the excited state is simply the sum of the individual components such that

$$\frac{dA^*}{dt} = -(k_F + k_{nr} + k_{ISC} + k_Q[Q])[A^*] \tag{15.8}$$

and thus by Equation 13.22

$$[A^*] = [A_0]e^{-(k_F+k_{nr}+k_{ISC}+k_Q[Q])t}. \tag{15.9}$$

Recall from Chapter 13 that if we consider two parallel reactions

$$A \overset{k_B}{\rightarrow} B$$

$$A \overset{k_C}{\rightarrow} C,$$

then the rate of disappearance of A is

$$-\frac{d[A]}{dt} = (k_B + k_C)[A],$$

thus the concentration of A disappears exponentially

$$[A] = [A]_0 e^{-(k_B+k_C)t}$$

Correspondingly, the rates of appearance of B and C are

$$\frac{d[B]}{dt} = k_B[A] = k_B[A]_0 e^{-(k_B+k_C)t}$$

$$\frac{d[C]}{dt} = k_C[A] = k_C[A]_0 e^{-(k_B+k_C)t},$$

which yields

$$[B] = k_B \frac{[A]_0}{(k_B + k_C)} \left\{ 1 - e^{-(k_B+k_C)} \right\}$$

$$[C] = k_C \frac{[A]_0}{(k_B + k_C)} \left\{ 1 - e^{-(k_B+k_C)} \right\}.$$

The consequence of this analysis is that the two products B and C appear in a constant ratio that depends only on the relative rate constants

$$\frac{[B]}{[C]} = \frac{k_B}{k_C}.$$

It follows that the fractional yields of each of the two species are

$$\frac{[B]}{[B] + [C]} = \frac{k_B}{k_B + k_C}$$

$$\frac{[C]}{[B] + [C]} = \frac{k_C}{k_B + k_C}.$$

15.2.1.3 Quantum yield of fluorescence

In the field of fluorescence spectroscopy, the primary measurement is the yield of photons emitted in the fluorescence process. The quantum yield of fluorescence is defined as the ratio of the number of photons emitted from the excited state to the number of photons absorbed to create the excited state. The former will be related to the rate constant for fluorescence, k_F, and the latter will be related to the sum of all the rate constants for all of the processes by which the excited state can decay, here $(k_F + k_{nr} + k_{ICS} + k_Q[Q])$. Thus, the quantum yield of fluorescence, Φ_F, is

$$\Phi_F = \frac{k_F}{(k_F + k_{nr} + k_{ICS} + k_Q[Q])} = \frac{k_F}{\sum_i k_i}. \tag{15.10}$$

We can note that the quenching process depends on the concentration of the quencher, $[Q]$, so it is possible to compare the fluorescence quantum yield as a function of the concentration of the quencher with that in its absence. Specifically, if the quencher is absent, $[Q] = 0$, the quantum yield is

$$\Phi_F^0 = \frac{k_F}{(k_F + k_{nr} + k_{ICS})}. \tag{15.11}$$

This means that the ratio of the quantum yield in the absence of the quencher to the quantum yield in the presence of the quencher becomes

$$\frac{\Phi_F^0}{\Phi_F} = \frac{k_F}{(k_F + k_{nr} + k_{ICS})} \times \frac{(k_F + k_{nr} + k_{ICS} + k_Q[Q])}{k_F} = 1 + k_Q'[Q], \tag{15.12}$$

which means that a plot of the inverse of the quantum yield of fluorescence as a function of a quencher should be a linear function of the quencher. This type of plot is called a Stern–Volmer plot as illustrated schematically in Figure 15.10.

It is interesting to note that there are two extreme forms of quenching: **dynamic and static quenching**.

Dynamic quenching In the first of these, the quenchers interact dynamically with all the molecules in their excited states and form a truly competing pathway for relaxation as indicated in the kinetic analysis of Equation 15.12 that leads to the Stern–Volmer analysis. In this case, the *lifetime of the excited state is also decreased*.

In the absence of the quencher

$$\left(\frac{dA^*}{dt}\right) = -k_T [A^*]^0 = -(k_F + k_{nr} + k_{ISC})[A^*]^0, \tag{15.13}$$

which leads to an exponential decay with a characteristic time constant of τ_0,

$$[A^*]^0 = [A^*]_0 \, e^{-(k_F + k_{nr} + k_{ISC})t} = [A^*]_0 \, e^{-t/\tau_0}. \tag{15.14}$$

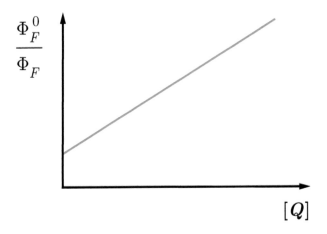

Figure 15.10: Illustration of the Stern–Volmer plot.

In the presence of the quencher

$$\left(\frac{dA^*}{dt}\right) = -k_T\,[A^*]^Q = -\left(k_F + k_{nr} + k_{ISC} + k_Q\,[Q]\right)[A^*]^Q, \qquad (15.15)$$

which leads to an exponential decay with a characteristic time constant of τ,

$$[A^*]^Q = [A^*]_0\,e^{-(k_F + k_{nr} + k_{ISC} + k_Q[Q])t} = [A^*]_0\,e^{-t/\tau_Q}. \qquad (15.16)$$

This yields a relation similar to the Stern–Volmer relation for the ratio of the lifetimes as

$$\frac{\tau_0}{\tau_Q} = 1 + k'_Q\,[Q]. \qquad (15.17)$$

Thus dynamic quenching is characterized by a linear dependence of both the quantum yield and the lifetimes of fluorescence decay on the quencher concentration.

Static quenching In the second case, the quenchers interact only with a subset of the molecules in the excited state, but when they do, they quench them completely; they effectively become invisible to the fluorescence measurement. As a consequence, the fluorescence quantum yield will also decrease linearly with the quencher concentration as more and more molecules are affected and we expect likewise that Equation 15.12 will hold.

On the other hand, the fluorescence now arises only from unaffected molecules, whose lifetimes are therefore not affected by the presence of the

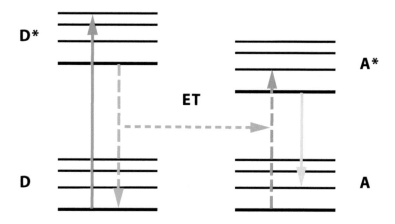

Figure 15.11: The energy transfer process.

quencher. Accordingly, the lifetime of the measured fluorescence will be *independent of the quencher concentration*.

Thus static quenching is characterized by a linear dependence of only the quantum yield of fluorescence on the quencher concentration.

Distinction between dynamic and static quenching therefore generally requires measurement of the lifetime of fluorescence.

15.2.2 Förster or fluorescence resonance energy transfer

Energy transfer refers to the process whereby the relaxation of one molecule causes the excitation of another molecule without emission or absorption of light. This process requires spin conservation and is therefore usually from singlet to singlet. The process is schematically illustrated in Figure 15.11.

Following the excitation from the ground state D to the excited state D^*, the interaction of D^* with the ground state A leads to an energy transfer that causes a simultaneous transition from D^* to D and from A to A^*. Subsequently, the excited state A^* can decay by the usual processes available such as non-radiative processes or emission of a photon as a fluorescence. In this energy transfer process, D is called the **donor** while A is called the **acceptor**.

15.2.2.1 Mechanisms of energy transfer

From the perspective of the donor, the energy transfer process is another form of quenching, and the quantum yield of the donor is decreased accordingly. There are different possible mechanisms for energy transfer.

If the donor-acceptor pairs are very efficiently coupled, the energy transfer is 100 % effective for those molecules in contact, and as with static quench-ing, the quantum yield, but not the lifetime of the donor, is affected by the presence of the acceptor. This is explained by the **Dexter energy trans-fer** mechanism that requires orbital overlap that allows simultaneous electron transfers from the excited state of the donor to the excited state of the accep-tor and from the ground state of the acceptor to the ground state of the donor. This mechanism is also called **Dexter electron transfer**. The Dexter elec-tron transfer mechanism may also be responsible for many static quenching mechanisms and usually requires molecular proximity of 1 nm or less.

Because the Dexter mechanism involves an electron exchange, it can be from a singlet donor to create a singlet acceptor, but it can also be from a triplet donor to create a triplet acceptor. In addition, it is possible for two triplet states to exchange electrons to form two singlets. This is termed **triplet-triplet annihilation**.

Triple-triplet annihilation is the mechanism whereby dioxygen (or molecular oxygen, O_2), whose ground state is a triplet, quenches excited state triplets of many molecules and in turn creates singlet oxygen. Sin-glet oxygen is very reactive and can oxidize a number of unsaturated molecules, or form other reactive oxygen species such as hydroxyl radi-cals ($\cdot OH$) or the superoxide anion (O_2^-). These processes are utilized in **photodynamic therapy**. This involves injecting or administering pho-toactive agents into tissues and irradiating these tissues. The photoactive agents are chosen for their effective intersystem crossing processes so that they form the triplet in high yield. Oxygen in the tissue will then inter-act with this triplet and, through the triplet-triplet annihilation process, form high local concentrations of singlet oxygen and other reactive oxy-gen species. These can in turn oxidize vital cellular molecules, for example nucleic acids, and cause cell death (apotosis). Photodynamic therapy has been useful to treat some cancers, particularly in areas where irradiation with visible or ultraviolet light is possible.

Förster proposed an alternative mechanism for energy transfer for molecules that are in close proximity (2-10 nm) but not in molecular contact. This involves interaction, through space, of the transition dipole moments of the donor and the acceptor. This effect is sensitive both to the distance of separation and the relative orientation of the molecular transition dipoles as well as their magnitudes as measured through the absorption coefficient of the acceptor and the quantum yield of the donor.

Figure 15.12 illustrates the different processes.

- Panel A illustrates the Singlet-to-Singlet Förster Energy Transfer Mech-anism: $(^1D^* + {}^1A \rightarrow {}^1D + {}^1A^*)$.

- Panel B illustrates the Singlet-to-Singlet Dexter Energy Transfer Mech-anism: $(^1D^* + {}^1A \rightarrow {}^1D + {}^1A^*)$.

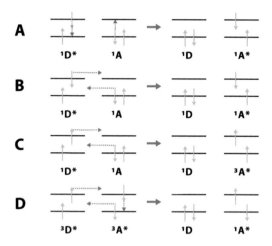

Figure 15.12: Simplified illustrations of the various energy transfer mechanisms. See text for legend.

- Panel C illustrates the Triplet-to-Triplet Dexter Energy Transfer Mechanism: $(^3D^* + {}^1A \rightarrow {}^1D + {}^3A^*)$.

- Panel D illustrates the Triplet-Triplet Annihilation Mechanism: $(^1D^* + {}^1A \rightarrow {}^1D + {}^1A^*)$.

15.2.2.2 Kinetics of energy transfer

The efficiency, and hence the rate, of energy transfer depends on the mechanism as well as a number of factors that depend on the relative orientation and separation of the donor and acceptor and on the energy match between excitation of the acceptor and relaxation of the donor. Accordingly, we would expect the rate constant for energy transfer, k_{ET}, to be generically given by

$$k_{ET} = f_1(spectral\ overlap) \times f_2(separation) \times f_3(orientation) \quad (15.18)$$

We will consider these in turn.

Overlap Integral   Irrespective of the mechanism of energy transfer, there must be an overlap of the energies that can lead to excitation of the acceptor and the energies that can cause relaxation in the donor. This requirement is embedded in the **overlap integral**, J_λ, which measures the extent to which the absorption spectrum of the acceptor, as measured by the extinction coefficient, $\epsilon_A(\lambda)$, and the emission spectrum of the donor, $F_D(\lambda)$, overlap. The overlap integral is given by their convolution

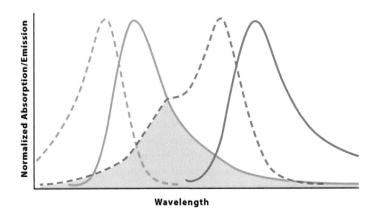

Figure 15.13: Illustration of the spectral overlap required to calculate the overlap integral J.the dashjed lines show the absorption spectra and the solid lines the emission spectra. The wavelengths in the overlap region are illustrated by the shaded area.

$$J_\lambda = \frac{\int_0^\infty \epsilon_A(\lambda) F_D(\lambda) \lambda^4 d\lambda}{\int_0^\infty F_D(\lambda) d\lambda}, \tag{15.19}$$

where the denominator ensures that the emission spectrum is normalized.

The overlap integral is a product of the emission and absorption spectra in the region where the emission and absorption spectra are both measurable as illustrated by the wavelengths corresponding to the shaded region in Figure 15.13.

Separation dependence   The Dexter mechanism is only effective at very short separations of the donor and acceptor and therefore requires a collision of the molecules so that they are in van der Waals contact. The probability of energy transfer decays exponentially as the separation, r, increases and hence

$$k_{ET}^{Dexter} = cJ_\lambda e^{-2r/d}, \tag{15.20}$$

where d is the sum of the van der Waals radii of the donor and acceptor and c is a constant. The exponential decay limits the Dexter mechanism to separations up to a few times the van der Waals radii, or 1-2 nm at most.

At larger separations, the energy transfer would have to be by the Förster mechanism, which involves interactions between the transition dipole moments through space. The potential energy associated with the dipole-dipole interaction is proportional to their separation to the inverse sixth power; this is the

Keesom interaction discussed briefly in Section 9.2. The interaction is assumed to be averaged over an ensemble of dipoles of donors, μ_D, and acceptors, μ_A and is given by

$$V = -\frac{2}{3(4\pi\epsilon)^2 kT}\frac{\mu_D^2\mu_A^2}{r^6}\kappa^2. \tag{15.21}$$

Here ϵ is the dielectric constant and κ is a function reflecting the relative orientation of the dipoles as discussed below.

The strength of the interaction between the transition dipole moments will depend on the spectral overlap integral, as discussed above. The rate constant for the energy transfer process will also depend on the quantum yield of the fluorescence , Φ_D, and the lifetime, τ_D of the donor. Thus

$$k_{ET}^{F\ddot{o}rster} = c'J_\lambda\frac{\Phi_D}{\tau_D}\frac{\kappa^2}{r^6}, \tag{15.22}$$

where c' is a collection of constants that include the refractive index,n, of the solution.

One can collect all the constants, including the quantum yield into a single constant, R_0, such that

$$k_{ET}^{F\ddot{o}rster} = \frac{1}{\tau_D}\left(\frac{R_0}{r}\right)^6 \tag{15.23}$$

where R_0 contains parameters that are specific for the donor-acceptor pair in question and is given by

$$R_0 = 8.79\times 10^2 \sqrt[6]{J_\lambda n^{-4}\Phi_D\kappa^2}. \tag{15.24}$$

[Here the constant reflects that R_0 is calculated in units of nanometers (nm)]

Orientation dependence The orientation of the colliding molecules is important in the Dexter mechanism since only specific relative orientations will allow the appropriate overlap of molecular orbitals. Other orientations will simply not result in energy transfer. The probability of a collision of the right orientation will be reflected in the constant, c, in Equation 15.20.

The relative orientation of the transition dipole moments of the donor and acceptor in the Förster mechanism is captured in the κ function. This is expressed in terms of four angles as illustrated in Figure 15.14: the angle ϕ between the two planes that contain the transition dipole moment vectors; the angle, θ_D, between the donor transition dipole moment and the vector, r, that connects the two transition dipole moment; and the angle, θ_A, between the acceptor transition dipole moment and r.

For two static molecules, the orientation factor, κ, is given by:

$$\kappa = cos\theta_T - 3cos\theta_D cos\theta_A \tag{15.25}$$

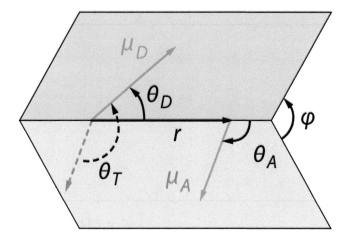

Figure 15.14: Defining the angles used to calculate the orientation factor κ.

with

$$cos\theta_T = sin\theta_D sin\theta_A cos\phi + cos\theta_D cos\theta_A. \qquad (15.26)$$

For molecules that are moving relative to each other, the κ factor is time-dependent, so it must be averaged over all the possible relative orientations of the dipoles during the lifetime of the donor and hence we generally need to calculate

$$\langle \kappa^2 \rangle_t = \left\langle (cos\theta_T(t) - 3cos\theta_D(t)cos\theta_A(t))^2 \right\rangle_t. \qquad (15.27)$$

Whenever the dipoles can move rapidly over all angles (isotropic motion), the average becomes $\kappa^2 = \frac{2}{3}$. This is often assumed when the donor and acceptor are on separate molecules, but is not always the case when they are on the same molecule, i.e., if there is intramolecular energy transfer. If there is a random relative orientation of stationary molecules, then the orientation factor is averaged over all orientations and is close to $\kappa^2 = 0.47$.

Energy transfer efficiency   The energy transfer efficiency, E, is defined as the fraction of the photons absorbed by the donor that result in energy transfer from the donor to the acceptor, so it is effectively the "quantum yield of energy transfer." In analogy with other quantum yield calculations we therefore get, using Equations 15.10 and 15.23

$$E = \frac{k_{ET}}{(k_{nr} + k_F + k_{ISC}) + k_{ET}} = \frac{k_{ET}}{1/\tau_D + k_{ET}} = \frac{1}{1 + \left(\frac{r}{R_0}\right)^6}. \qquad (15.28)$$

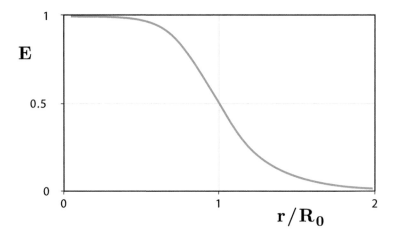

Figure 15.15: The dependence of the energy transfer efficiency on the separation of the transition dipole moments.

This shows us that there is a relatively sharp distance dependence of the efficiency of energy transfer process as illustrated in Figure 15.15.

The strong distance dependence of the energy transfer efficiency means that it can be used to determine the distance between two transition dipole moments reasonably accurately whenever they are within the range of about $0.5 < r/R_0 < 1.5$. The value of R_0 is dependent on the particular donor-acceptor pair, but typically R_0 ranges from one to as many as ten nanometers. This means that fluorescence energy transfer can be a useful tool to determine the size or shape of nanoscale structures and has proven very useful in the study of structural relationships in biological systems, particularly shapes and sizes of macromolecules.

15.2.2.3 Measurements of energy transfer

There are a number of approaches to measuring the energy transfer process, specifically the energy transfer efficiency. Since the process is, from the perspective of the donor, a quenching process, the best approach is to measure the donor fluorescence quantum yield as a function of the concentration of the acceptor. This is frequently not possible, particularly when the energy transfer is between two molecules on the same nanoparticle (or macromolecule). Still, measuring the fluorescence yield of the donor in the absence, F_D, and the presence, F_{DA}, of the acceptor one can calculate the energy transfer efficiency as

$$E = \frac{F_D - F_{DA}}{F_D} = 1 - \frac{F_{DA}}{F_D}. \tag{15.29}$$

We saw earlier that the lifetimes of fluorescence of the donor can be used to measure the efficiency of a quenching process. Correspondingly, the energy transfer efficiency can be calculated from the lifetime in the absence, τ_D, and the presence, τ_{DA}, of the acceptor as

$$E = \frac{\tau_D - \tau_{DA}}{\tau_D} = 1 - \frac{\tau_{DA}}{\tau_D}. \tag{15.30}$$

In many cases, the most effective way to measure the lifetime of fluorescence in the absence of the acceptor is to remove it by photochemically destroying it — or photobleaching it — and measuring the increase in donor fluorescence afterward.

CHAPTER 16

Application of fluorescence microscopy

The use of fluorescence spectroscopy is widespread and is not specifically aimed at the nanoscale world. However, the combination of fluorescence tools with microscopy tools has proven extremely useful in the domain of nanobiotechnology where the fluorescence provides exquisite sensitivity and the microscopy provides high spatial resolution allowing studies of individual molecules or nanoparticles and their behavior.

The purpose of this chapter is to illustrate the power of this combination of tools to study selected biological systems.

16.1   FLUORESCENCE MICROSCOPY

The fluorescence microscope derives from the optical microscope because the majority of fluorescence properties are tied to optical transitions in organic type molecules.

16.1.1   Microscope optics

Most modern optical microscopes have infinity corrected optics as illustrated schematically in Figure 16.1.

The optical path is constructed such that there are parallel rays between the objective lens and the tube lens, which allows these to be moved relative to each other and for optical elements, such as filters, to be placed between them without changing the focal properties of the microscope. There is an intermediate image plane which acts as the object plane for the eyepiece and the parallel rays beyond them are ideal for observation by the human eye. The intermediate image plane can also be projected directly onto a camera or another detector system.

The epifluorescence microscope takes advantage of the infinity space be-

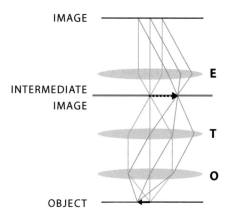

Figure 16.1: Schematic view of infinity corrected microscope optics. The objective lens (O) and the tube lens (T) allow parallel rays between them and ensures parallel rays beyond the eyepiece (E).

tween the objective lens and the tube lens by placing a dichroic[1] mirror between them at a forty-five-degree angle as illustrated in Figure 16.2. The dichroic mirror is designed such that light below a particular wavelength is reflected off the surface while light above that wavelength is transmitted. The light from a light source is brought onto the dichroic mirror after being filtered through an excitation filter that transmits only short wavelength light[2] that will be reflected by the dichroic mirror (the blue light path in Figure 16.2). Subsequently, the reflected light is focused by the objective lens onto the specimen in the object plane.

Fluorescence light emitted at longer wavelengths will be collected by the objective lens (the green light path in Figure 16.2) and transmitted straight through the dichroic mirror and focused by the tube lens onto the image plane. Any light at the excitation wavelength being reflected back from the specimen will be reflected at the dichroic mirror, effectively returning along the path by which it arrived. However, since the dichroic mirror is not perfect, and because the intensity of fluorescence can be weak relative to the reflected light, the light is further filtered by an emission filter that only transmits long wavelength light[3] also transmitted by the dichroic mirror.

Depending on the illumination source and the focusing optics on the excitation side, it is possible to illuminate the entire sample in the field of view of the objective at the object plane or to illuminate a small region only. For example, if the source is a laser, the laser can be focused to a single small spot

---

[1] dichroic means two-color
[2] called a low pass filter
[3] called a high pass filter

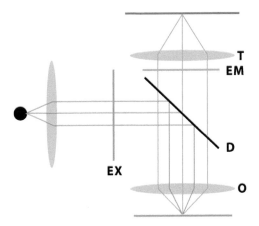

Figure 16.2: Schematic view of the optical path in an epifluorescence micro-scope. A combination of an excitation filter (EX), a dichroic mirror (D) and an emission filter (EM) is placed in the infinity space between the objective lens (O) and the tube lens (T).

on the specimen. Correspondingly, if the excitation light is caused to fill the back of the objective with a uniform light field, the light will be focused to a diffraction limited spot at the center of the optical field in the object plane. These techniques create a focal spot in a thin specimen and it is therefore possible to determine the fluorescence arising from a small region of the specimen. This does not, however, generate an image of the specimen.

To obtain an image, the whole field of view is illuminated evenly and the fluorescence arising from different parts of the specimen are focused on different parts of the image plane. A sensitive camera, typically a CCD Camera, is then placed in the image plane to capture the fluorescence emission image.

A CCD camera (or charge coupled device camera) consists of a two-dimensional array of photosensitive elements that capture the photons and emit an electron (by the photoelectric effect discussed in Chapter 5) that is captured by a capacitor where it is stored. The accumulation of charge on the capacitor is a measure of the number of photons that hits the photosensitive element and when it is read out provides a measure of the intensity of emission from that particular region of the sample. CCD cameras are also used extensively in modern digital cameras and they can be very sensitive.

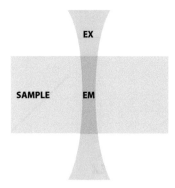

Figure 16.3: Illustration of the focus of the excitation light.

16.1.2 Confocal microscopy

While it is possible to focus the excitation to a small region of space in the object (or sample) plane, the excitation beam will penetrate the whole sample, and unless it is very thin, the emission of fluorescence will occur from regions above and below the focal plane as well. The volume from which emission can occur is not as small as would be desired, particularly along the direction of the illumination (the z-direction, if the focal plane is considered to be in the x,y-plane). This is illustrated in Figure 16.3.

It is evident that emission will occur throughout the sample wherever the excitation light penetrates.

In order to minimize the total volume, and in particular the direction along the z-axis, from which emission is **detected**, it is possible to introduce a pinhole in the intermediate image plane of the microscope. The effect of this is to eliminate the emission that does not arise from the region immediately above and below the focal plane since that emission will not be focused on the pinhole in the image plane. This is illustrated in Figure 16.4. The consequence is that the vertical extent of the detection volume is reduced and the overall effective volume arising from the combination of a focused excitation beam and a focused emission detection becomes an ovoid volume as illustrated in Figure 16.5. This is called the **confocal volume** and the microscope is called a **confocal microscope**.

While the confocal volume is a well-defined volume centered at the intersection of the focal plane and the vertical axis of the microscope, it is a single volume in the sample. In order to generate an image of the sample, it becomes necessary to scan the sample in the x,y-plane, or to scan the excitation spot across the sample. The latter is the more common and faster approach and most confocal microscopes use a laser[4] as the light source and mirrors in the

---

[4]The laser serves the dual function of a very narrow excitation wavelength and a point

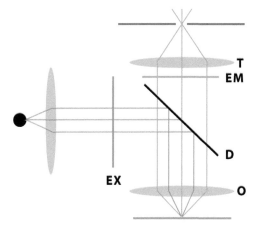

Figure 16.4: Illustration of the effect of a pinhole in the image plane.

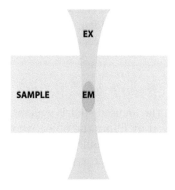

Figure 16.5: Illustration of the confocal volume.

microscope that can change the location of the excitation volume. This type of microscope is called a laser scanning confocal microscope.

16.1.2.1   The shape of the confocal volume

The shape in the x,y-plane   A laser is generated in an optical cavity and the light that is emitted is normal controlled to be in what is called the TEM_{00} mode, which means that the cross-sectional intensity profile has a Gaussian distribution. If this profile is faithfully transmitted through the lenses of the

---

source, but the latter can also be achieved by illumination through an optical fiber. The combination of a laser delivered through an optical fiber provides optimal flexibility in design.

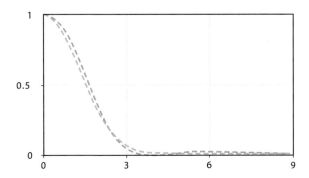

Figure 16.6: Comparison of the Airy disc intensity profile (blue) with the equivalent Gaussian profile (orange).

microscope, the excitation profile in the x,y-plane is a two-dimensional Gaussian function characterized by a width, w_0, which corresponds to the width of the intensity profile along any direction where the intensity is e^{-2} of the central intensity, I_0

$$I(x, y) = I_0 e^{-2(x^2+y^2)/w_0^2}. \tag{16.1}$$

If, on the other hand, the focused spot is generated by a diffraction limited aperture (which is achieved by filling the objective with uniform illumination) then the cross-sectional intensity profile is that of the airy-disc pattern. This is given along any direction, r, in the x,y-plane by

$$I(r) = I_0 \left(\frac{2J_1(r)}{r}\right)^2. \tag{16.2}$$

Here $J_1(r)$ is the Bessel function of the first kind of order one (which we also encountered in the solution to the electron in a circular disc in Chapter 4) and $r = ka\sin\theta$ where k is the wavenumber, a is the radius of the aperture, and θ is the angle away from the direction perpendicular to the aperture.

It turns out that this intensity profile is also approximated quite well by a Gaussian function. This is illustrated in Figure 16.6.

For most practical purposes, it is therefore possible to consider the intensity distribution of the confocal volume anywhere in the x,y-plane to be Gaussian as in Equation 16.1. The parameter, w_0 is called the beam width.

The shape along the z-direction Since the intensity profile along the z-direction is effectively determined by the size of the aperture in the image plane, to a first approximation, this intensity profile can also be characterized by a Gaussian distribution. However, the distribution is much broader and

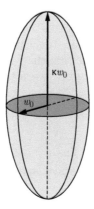

Figure 16.7: The confocal volume.

is characterized by a width $w_z = \kappa w_0$. The κ is a fixed multiplication factor that depends on the size of the pinhole aperture and is usually determined experimentally. It typically ranges from around 4 to 15.

Thus, the intensity profile of the confocal volume is reasonably well characterized by a three-dimensional Gaussian function

$$I(x, y, z) = I_0 e^{-2(x^2+y^2)/w_0^2} e^{-2z^2/\kappa^2 w_0^2}. \tag{16.3}$$

This is also called the **point-spread function** for the confocal volume.

The confocal volume is shown in Figure 16.7 as a contour where the intensity has a value of e^{-2} relative to the value at the center of the volume. This represents a region of space that encloses about 70% of the molecules that can be detected, thus the confocal volume, $V_{confocal}$ is smaller than the effective observation volume $V_{effective}$.

The confocal volume is given by

$$V_{confocal} = \left(\frac{\pi}{2}\right)^{3/2} w_0^3 \kappa = \left(\frac{1}{2}\right)^{3/2} V_{effective}. \tag{16.4}$$

16.1.2.2   The size of the confocal volume

The size of the confocal volume will depend on how it is created. If the source of illumination is a laser, the beam width at the focal plane, w_0, is controlled by the size of the laser beam at the back image plane and the magnification of the objective lens. Typically, the beam width can be adjusted from a fraction of a micrometer to a few micrometers as desired. If on the other hand the illumination is designed to fill the back of the objective the beam width will be diffraction limited.

Diffraction limitation and Abbe's law   The diffraction limit refers to the smallest possible distance that can be resolved when observed with electromagnetic radiation of a particular wavelength. The diffraction limit, d, in a microscope depends on three factors: the wavelength of the radiation, λ; the refractive index, n, of the medium between the sample and the objective; and the angle, 2θ, subtended by the objective lens as described by the Abbe formula

$$d = \frac{\lambda}{2n\sin\theta} = \frac{\lambda}{2NA}, \tag{16.5}$$

where $NA = n\sin\theta$ is the numerical aperture of the objective lens of the microscope.

The diffraction limit is reduced when the wavelength is small and when the numerical aperture is large. The numerical aperture is often improved by increasing the refractive index using special oil as the medium and by designing the lenses to operate at short focal distances where the subtended angle is larger. Typical numerical apertures for high resolution fluorescence microscopy range from $NA = 1.4$ to 1.6, which for wavelengths around 500 nm will, in principle, give diffraction limits as small as $d = 500/(2 \times 1.6)$ nm $\cong$ 150 nm. In practice, it is difficult to attain resolutions beyond 200 nm, however. The confocal microscope is therefore not of sufficiently high resolution to resolve nanoscale objects.

Based on the Abbe diffraction limit, the smallest confocal volume that can be expected in a confocal microscope at the highest resolutions are therefore on the order 0.16 μm^3 or 0.16 fL, corresponding to an effective volume of about 0.45 fL. This suggests that, while it is not possible to resolve nanoscale structures, it is possible to study individual nanoparticles when the concentration is at the nanomolar level. Specifically, in a $3 - nM$ solution the average number of molecules in the smallest confocal volume would be $n = 3 \times 10^{-9} \times 6.02 \times 10^{+23} \times 0.45 \times 10^{-15} \cong 1$.

Correspondingly, the smallest area that can be observed is on the order of 0.13 μm^2, which means that it is possible to study individual nanoparticles on a surface if the density is less than about 7 μm^{-2}.

At densities or concentrations low enough that only single particles are observed at any given position or time, it becomes possible to calculate the position of the nanoparticle at much higher precision than expected from the Abbe diffraction limit.

This possibility arises particularly in fluorescence microscopy if we can sample the profile of the fluorescence emission intensity at a high precision, in other words, if we sample the intensity across the confocal volume. This requires moving the particle within the confocal volume (or the confocal volume around the particle) in smaller steps than the size of the confocal volume. For example, if the width of the confocal volume is characterized by a $w_0 = 200\ nm$, then, if the confocal volume can be sampled with $10 - nm$ steps in the x, y-plane and $100 - nm$ steps in the z-direction, there will be about 10 measures of the intensity profile in each direction. Given that the shape of the intensity profile – the point spread function – is known, as in Equation 16.3, it is possible to calculate the location of the particle where the intensity is maximal with a precision much greater than the diffraction limit. This concept has been known for some time and led to the development of one type of super-resolution microscopy.

16.1.3 Super-resolution microscopy

Super-resolution microscopy refers to a set of approaches of creating 2D or 3D images of fluorescent objects with a resolution that exceeds that expected by the Abbe diffraction limit. It may also be considered as sub-diffraction microscopy. In general terms, the approaches to enhancing the resolution relies on exploiting some specific aspects of optics or on relying on observing only a single molecule or particle at a time.

16.1.3.1 Optical techniques

There are some techniques that rely on applying optical confinement or optical interference effects. These are by some considered true super-resolution techniques and include near-field optical microscopy, 4Pi microscopy, and structured illumination microscopy.

Near-field scanning optical microscopy   One of the first approaches to improve the resolution in imaging was to confine the molecules of interest to the volume close to an aperture with dimensions smaller than the diffraction limit and letting the excitation light emerge through this aperture. The region immediately around the aperture at distances much less than the wavelength of the light is called the **near-field** and molecules in that volume can be excited and fluoresce. The detection can be confined to the same region, either using a microscope objective or detecting the emission through the same aperture and the resolution is now effectively determined by the size of the aperture. The principle is illustrated in Figure 16.8.

The excitation light is confined to a metal coated optical fiber which has an aperture at the end of a tapered section. The excitation light will emerge and in the far field will diverge as indicated. Close to the aperture there is a narrow region, the near-field where the dimension of the excitation light is comparable to the size of the aperture. Samples in this region will now be measured with a resolution defined by the size of the aperture, typically 50 to 100 nm.

In order to create an image, the aperture needs to be scanned across the sample in a raster fashion in steps less than or comparable to the size of the aperture. Because the near field is restricted to very small distances from the aperture, this type of microscopy is sensitive only to the molecules on the surface of the sample. However, since the tip must remain close to the

Figure 16.8: The near-field aperture and volume.

surface throughout, the scanning process will also provide a measure of the topography of the sample since the tip must be moved up or down whenever the height of the surface changes.

4Pi microscopy The resolution in the confocal microscope is much worse along the z-direction than in the x, y-plane by as much as a factor of 10 because it is controlled by the image plane pinhole. The resolution along the z-direction can be improved using 4Pi microscopy, in which the sample is contained between two objective lenses. The excitation light is split and passed through both objectives and caused to be super-positioned in a small volume at the common focal point. The fluorescence emitted from molecules in this small volume is collected by both objectives and combined in the emission path. The resolution along the z-direction can be improved by as much as a factor of 7, depending on the detailed operation of the microscope, to as little as $150 - 200\ nm$.

Structured Illumination Microscopy The limitation of resolution in the spatial dimension observed in the microscope corresponds to the fact that there is no information in reciprocal space at spatial frequencies above a threshold frequency given by the inverse of the diffraction limit $k_{max} = 2NA/\lambda$. Imposing a structured illumination, such as a cosine modulation of the otherwise even illumination, creates an interference between the structured illumination pattern and the sample patterns that introduces both difference and sum frequencies. The sum frequencies extend the range of frequencies that can be observed and hence improves the resolution by as much as a factor of two in the direction of the structured illumination. An example of the interference pattern is illustrated in Figure 16.9 by two periodic patterns overlapping at an acute angle. By applying the structured illumination in known orientations, it is possible to improve the resolution in all directions by a factor of two.

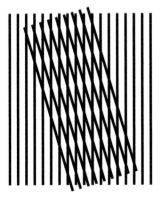

Figure 16.9: Illustration of the interference patterns arising from overlapping periodic patterns.

There are variations of the illumination pattern that in principle can improve the resolution by more than a factor of two and that allow enhanced resolution along the z-direction as well.

16.1.3.2 Stimulated emission or saturation techniques

There are other techniques that rely on manipulating the populations in the excited or ground states in a spatially controlled fashion that result in the effective observation volume being reduced. These include Stimulated Emission Depletion Microscopy, Ground State Depletion Microscopy, and Saturated Structured Illumination Microscopy.

Stimulated Emission Depletion Microscopy When the molecules in a region defined by the cross-section of a laser beam focused in a microscope are excited by a short pulse, the population in the excited state is distributed in space in proportion to the laser intensity in that region of space. They will therefore be distributed as a two-dimensional Gaussian function in the x, y-plane with a width defined by the cross-section of the point spread function (Equation 16.3), which may be diffraction limited. The Gaussian shape arises from the use of the TEM_{00} mode of the laser cavity. Higher order modes of the laser can be used to form a toroidal cross-section (effectively the shape of a doughnut) of intensity where there is no intensity at the center of the beam.

The transverse electromagnetic modes (TEM) of the laser beam are modes in which there are no electric or magnetic fields along the direction of propagation. They are products of a Gaussian function and La-

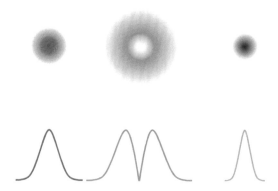

Figure 16.10: Illustration of the stimulated emission depletion on the size of the emission region.

> guerre polynomials and represent solutions to the electromagnetic fields in a cylindrical space, somewhat analogous to the solutions to the wave function in a disc as solved in Section 4.2.3.2.

If the region of space containing the excited state molecules is illuminated with a second pulse with this toroidal shape, the excited molecules can be caused to return to the ground state by stimulated emission, thereby decreasing the number of molecules in the exited state in the regions away from the center of the beam. This is termed stimulated emission depletion. The net result is that subsequent fluorescence emission will arise only from those molecules that remain excited in the central portion of the beam. This region is smaller than the diffraction limited beam, and hence the effective resolution is improved. The concept is illustrated in Figure 16.10.

Here the excitation region is shown at the left as a Gaussian shaped, diffraction limited spot. The toroidally shaped stimulated emission is shown in the center and the result is a smaller region, shown at the right, from which the remaining excited molecules can emit fluorescence. The width of the emission region is much smaller than the original excitation region and hence the resolution is increased significantly. In fact, the final resolution can be determined by a modification of Abbe's equation

$$d = \frac{\lambda}{2NA\sqrt{1 + \frac{I_{max}}{I_{sat}}}}. \tag{16.6}$$

Here, I_{max} is the peak intensity of the laser beam that causes stimulated emission and I_{sat} is the intensity needed to cause saturated emission depletion by a factor of $1/e$.[5] Whenever $I_{max} > 0$ the resolution is enhanced, but not significantly unless $I_{max} >> I_{sat}$.

The principle of STED microscopy is based on the ability to cause specific molecules in a defined region of space to become effectively invisible – they are placed in an "off" state during the observation period by being returned to the ground state.

Ground state depletion microscopy   As an alternative to turning the molecules off by returning them to the ground state as in the STED microscopy, it is possible to stimulate them to transition to the triplet state, where they remain while the fluorescence is observed because of the longer lifetime of the triplet state. If this is achieved in the same geometrical structure as in STED, the effect is an equivalent resolution enhancement as shown in Equation 16.6.

Saturated structured illumination microscopy   Saturated structured illumination combines the principles of structured illumination as described above with the ability to saturate some of the fluorescent molecules to create additional interference fringes thereby enhancing the resolution beyond the factor-of-two limitation in normal structured illumination microscopy.

While there is, in principle, no limit to the resolution that can be achieved by these three techniques, there are limitations arising from the time to collect the information to achieve an image. For STED and GDS and analogous techniques, the measurements are in a small region at a time, so it is necessary to scan the observation region over the sample to generate an image, which can limit the temporal resolution of an experiment. While the structured illumination excites across the entire image, it must be done repeatedly with different pattern orientations, which also slows down the experimental acquisition. In all cases, the fluorescent molecules must be relatively photostable since they are being exposed to very large excitation intensities.

16.1.3.3   Localization techniques

In contrast to the physical manipulation of the light path or the regions of excitation and emission described above, there are a number of techniques that rely on creating conditions where there is a high probability that only one molecule is excited and emitting fluorescence in any given region. These conditions then permit calculation of the position of that molecule with high accuracy whenever the point-spread function is known with high precision (Equation 16.3). The accuracy of localization of the molecule to the central point in the point-spread function will depend on how accurately the emission

---

[5]This is based on a first order process of emission depletion by the light that competes with the natural emission process and is therefore inversely dependent on the emission lifetime.

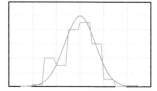

Figure 16.11: Illustration of the accuracy of localizing a molecule at different photon intensity measurements. The left curve corresponds to poor photon statistics, the right to good photon statistics.

is determined as a function of space. This in turn will depend on how many photons can be collected from each molecule during the observation. This concept is illustrated in one dimension in Figure 16.11.

This figure illustrates the fraction of photons collected in a set of discrete channels of fixed dimension (orange curves) relative to the expected point-spread function. The more photons are collected, the better the distribution approximates the ideal point-spread function and the better the determination of the location of the molecule. Clearly, the resolution is ultimately limited by the effective dimension of the detector channel. Still, with these approaches, it is possible to achieve localization accuracies of about 10 nm or better.

There are a number of approaches to ensuring that only single molecules are observed in each region at any given time. Most of these rely on photo-chemically switching molecules on or off in a stochastic manner.

There are variations on the principle that differ in subtle ways and these methods include Stochastic Optical Reconstruction Microscopy (STORM) and Photo Activated Localization Microscopy (PALM). In both cases, the fluorescent molecules are switched from a dark state to a fluorescent state with appropriate wavelength lasers at a rate low enough to ensure that only a small fraction of molecules are observed. Once observed, they may be switched back to a dark state or photochemically destroyed (bleached) so that they are not observed again. While the STORM approach works well with a number of fluorophores that can be caused to blink, the PALM technique relies on a photo-activatable probe as the name implies. One advantage of the latter is that it is in principle possible to perform quantitative estimates of the total number of fluorescent molecules.

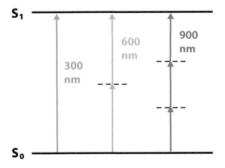

Figure 16.12: Illustration of the concept of multi-photon excitation.

16.1.3.4 Multi-photon techniques

In the discussion of excitation of molecules from the ground state to the excited state, we have so far only considered the excitation arising from a single photon or the appropriate energy of wavelength. However, it is possible for two or more photons of lower energies (and longer wavelengths) to simultaneously interact with a molecule and thereby effect the same transition. This concept of multi-photon transitions is illustrated in Figure 16.12.

The possibility exists that a transition that normally would require a high energy at a wavelength of 300 nm can be achieved by application of two photons with half the energy at 600 nm or by three photons with a third of the energy at 900 nm. Clearly, the probability that two or three or more photons arrive at the same place at the same time increases if the photon flux is high, i.e., as the intensity or the power of the applied light increases. Typically, these intensities are not present under normal illumination from a laser with continuous output – a cw laser. However, if the output is condensed into short pulses, the peak powers or intensities can rise dramatically and will enable multi-photon transitions. For a laser delivering pico-second pulses at a frequency of about 1000 per second, for example, the peak power (intensity) will be about 10^9 times the average power (intensity).

Multi-photon spectroscopy can be useful since it allows illumination with a lower energy, longer wavelength, source which can have a number of advantages. For example, glass does not transmit light in the ultraviolet, so optical microscopes cannot be used in the ultraviolet, but illumination with a high power pulse in the visible can access these high energy transitions in a two-photon transition and the longer wavelength emission may then be observable in the normal microscope.

Similarly, illumination with pulses in the infra-red can access transitions in the visible with at two-photon transition allowing for illumination with energies where there is less damage to tissues or cells. Also, some tissue is more transparent in the infrared region of the spectrum, so the illumination can penetrate further and this can facilitate fluorescence measurements in otherwise inaccessible parts of the tissue.

Most importantly, since the multi-photon absorption is very sensitive to the power of the light, the probability of absorption falls off very quickly and the excitation is therefore automatically confined to a rather small volume in space, defining a three-dimensional point-spread function without the requirement of the emission pin-hole needed in the confocal microscope. In the case of two-photon microscopy with infrared light, the increase in the point-spread function arising from the longer wavelength is nearly compensated for by the excitation profile and the resolution is nearly the same as in the conventional confocal microscope.

However, with higher order multi-photon absorptions, the effective point-spread function will be defined by the very highest intensities at the center of the point-spread function and the effective volume within which the excitation occurs can become quite small. This concept has been exploited by using femto-second laser pulses and multi-photon absorption in materials, such as silica, to ablate the material exclusively at the central region of the laser to create nanoscale objects, i.e., using light pulses to perform nanomachining. An example of creating nanofluidic channels in this manner can be found in the work of Liao and coworkers (Optical Letters 38, 187 (2013)).

16.1.4 Single molecule FRET

Fluorescence resonance energy transfer is a useful tool to determine the separation or relative orientation of two fluorophores in space. The measurements are usually made in bulk and are therefore average measurements that reflect the overall distribution of these parameters among the population of molecules. There are times when it is important to understand what the underlying distribution looks like. For example, a measurement of energy transfer efficiency of 0.5 in a solution could be an average measure of either a single population in which all of the molecular pairs have energy transfer efficiencies of about 0.5, or of two populations, half of which have an energy transfer efficiency of 1 and the rest an energy transfer efficiency of 0. These represent very different molecular situations.

In order to determine the actual distribution of energy transfer values, one must measure these values for a sample of molecules, one at a time; this corresponds to single molecule FRET. The sensitivity of fluorescence in confocal microscopes makes this possible. The critical point is to measure on solutions that are sufficiently dilute that the probability of observing two molecules simultaneously is nearly zero, which in turn means that the frequency of observing the molecules is low.

Figure 16.13 shows two records of intensity of fluorescence measured as a function of time. One record is the fluorescence of the donor (green), the other

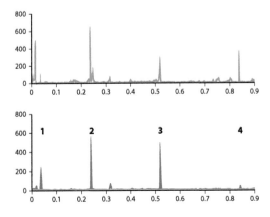

Figure 16.13: Measurements of single molecule energy transfer.

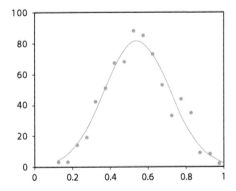

Figure 16.14: Example of the distribution of energy transfer efficiencies.

the fluorescence of the acceptor (red), **when only the donor is excited**. Each of the spikes observed in the record corresponds to a single fluorescent object entering and leaving the confocal observation volume, in the process absorbing and emitting light. Energy transfer is observed in the object or molecule in question whenever there is fluorescence measured in both channels simultaneously, and the efficiency of energy transfer can be calculated for each object or molecule from the intensities in each of the channels. Thus in this figure, the object measured at 1 exhibits a very large energy transfer efficiency since the emission in the acceptor channel is much larger than that in the donor channel. Correspondingly, the object at 2 has an energy transfer close to a half, that at 3 less than half, and that at 4 shows very little energy transfer.

When a large number of objects or molecules have been observed, it is pos-

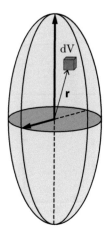

Figure 16.15: Location of a small volume in the confocal volume.

sible to build up a distribution of energy transfer measurements, as illustrated in Figure 16.14. In this case, the distribution is uniform, representing a single population of molecular entities.

16.2 FLUORESCENCE CORRELATION SPECTROSCOPY

In Section 13.2 we introduced the idea of using concentration fluctuations to measure dynamics, such as chemical kinetics or diffusion. In Section 13.4 we briefly proposed that the concentration fluctuations can be measured by fluctuations in a fluorescence signal. The purpose of this section is to expand on this concept, given the additional understanding of fluorescence spectroscopy and microscopy provided in Chapters 14 and 15 respectively.

Consider the measurement of a fluorescence emission signal from species i located in a volume dV at a distance $\mathbf{r}$ from the central point of a confocal volume as illustrated in Figure 16.15

The concentration of the species in that small volume is $c_i(\mathbf{r}, t)$ and will generally vary with position and time. The intensity of the illumination at that point in the confocal volume is $I(\mathbf{r})$ so we expect the fluorescence emitted, $dF_i'(\mathbf{r}, t)$, from that small volume to be:

$$dF_i'(\mathbf{r}, t) = g_i I(\mathbf{r}) c_i(\mathbf{r}, t) dV. \qquad (16.7)$$

Here, g_i is a parameter that accounts for the extinction coefficient and quantum yield of the species i of interest.

Since only a fraction of the photons emitted are detected, the fluorescence signal will depend on the collection efficiency, $f(r)$, of the confocal microscope, which also takes into account the efficiency of counting the photons that arrive at the detector. The detected fluorescence signal will then be

$dF_i(\mathbf{r}, t) = dF_i'(\mathbf{r}, t)f(\mathbf{r}, t)$. Integrating over the entire confocal volume gives the total measured signal, which is a function of time only.

$$F_i(t) = \int_{\infty}^{\infty} dF_i(\mathbf{r}, t)dV = \int_{-\infty}^{\infty} g_i f(\mathbf{r})I(r)c_i(\mathbf{r}, t)dV \qquad (16.8)$$

Fluctuations in the number of molecules in each of these small volumes, $\delta n_i(\mathbf{r}, t)$, correspond to fluctuations in concentrations $\delta c_i(\mathbf{r}, t)$, which lead to fluctuations in the fluorescence emitted and detected from each small volume. These fluctuations add to a temporal fluctuation in the total signal, $\delta F_i(t)$, given by

$$\delta F_i(t) = F_i(t) - \langle F_i(t) \rangle = g_i \int_{-\infty}^{\infty} f(\mathbf{r})I(r)\delta c_i(\mathbf{r}, t)dV, \qquad (16.9)$$

where $\delta c_i(\mathbf{r}, t) = c_i(\mathbf{r}, t) - \langle c_i(\mathbf{r}, t) \rangle$.

If there are multiple species contributing to the fluorescence measured at a given wavelength, the total fluctuation will reflect the sum of these signals

$$\delta F(t) = \sum_{i=1}^{M}(F_i(t) - \langle F_i(t) \rangle). \qquad (16.10)$$

As discussed in Chapter 13, the dynamic information of interest can be obtained by calculating the temporal autocorrelation function, $G(\tau)$, of the fluorescence signal, and the most general form of this then becomes

$$G(\tau) = \sum_{i=1}^{M}\sum_{j=1}^{N} \langle \delta F_i(t)\delta F_j(t+\tau) \rangle$$

$$= \sum_{i=1}^{M}\sum_{j=1}^{N} g_i g_j \int_{-\infty}^{\infty}\int_{-\infty}^{\infty} f(r)I(r)f(r')I(r') \langle \delta c_i(r, t)\delta c_j(r', t+\tau) \rangle \, dV dV'.$$

$$(16.11)$$

16.2.1 Sources of fluorescence fluctuations

There are primarily two general sources of fluctuations in the fluorescence signal that can contribute to the correlation function. First, the fluctuations can arise from **random noise** in the excitation or detection systems. If truly random, these are not correlated even at the smallest delay times, τ. These fluctuations will therefore only contribute to the correlation function at zero delay times, $G(0)$. However, since these noise fluctuations can be significant, their contribution to the amplitude at zero delay can be significant. This type of noise is called white noise or shot noise since it contributes to fluctuations at all frequencies. This noise does not affect the correlation function decay. Second, the fluorescence fluctuations can arise from **occupation number** *fluctuations*, which is what is of interest. There are three general ways in which the numbers can fluctuate:

1. The number of molecules can change through **diffusion** from one region of the confocal volume to another. In this case, the fluctuations are given by the Fick's second law

$$\frac{\partial \delta c_i(\mathbf{r}, t)}{\partial t} = D_i \nabla^2 \delta c_i(\mathbf{r}, t), \qquad (16.12)$$

   where D_i is the diffusion coefficient of species i.

2. The number of molecules can change through **convection** (flow) from one region of the confocal volume to another. In this case, the fluctuations are given by

$$\frac{\partial \delta c_i(\mathbf{r}, t)}{\partial t} = -V_i \frac{\partial c_i(\mathbf{r}, t)}{\partial x}, \qquad (16.13)$$

   where V_i is the velocity of flow along the x-direction (which is an arbitrary direction).

3. The number of molecules can change through **chemical reactions** within each of the small volumes. In this case, the fluctuations are given by

$$\frac{\partial \delta c_i(\mathbf{r}, t)}{\partial t} = \sum_{j-1}^{M} T_{ij} \delta c_i(\mathbf{r}, t), \qquad (16.14)$$

   where T_{ij} is the reaction rate matrix that represents all the possible rates of transformation of species i to another species j. For example, if i reacts at a rate of k_{ij} to form j the matrix element would be $T_{ij} = k_{ij} \langle c_i \rangle$.

It is, off course, possible that there could be a combination of these processes, in which case a combinations of these rate equations apply.

16.2.2 FCS of a single species

The solutions to Equation 16.11 for each of the above dynamic processes was first explored for a single species by Elson and co-workers for a geometry where the fluctuations would be dominated by the transverse motion in the x, y-plane because the variation in excitation intensity and collection efficiency varies more rapidly than any longitudinal motion along the direction of propagation. The problem then becomes effectively a two-dimensional problem with

$$f(r)I(r) = I_0 e^{\frac{-2\left(x^2 + y^2\right)}{w^2}} \qquad (16.15)$$

and the solutions they found were respectively.

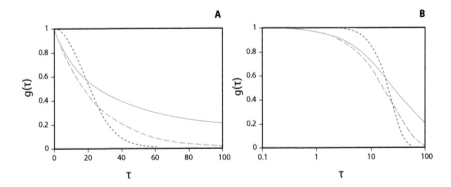

Figure 16.16: Typical autocorrelation function decay curves for diffusion (solid), convection (dotted), and reaction (dashed) processes (A) linear time scale and (B) logarithmic time scale.

1. For diffusion

$$G(\tau) = G(0)\frac{1}{1 + \frac{\tau}{\tau_D}}, \qquad (16.16)$$

   where $\tau_D = \frac{w^2}{4D}$ is the characteristic diffusion time.

2. For convection

$$G(\tau) = G(0)e^{-\left(\frac{\tau}{\tau_V}\right)^2}, \qquad (16.17)$$

   where $\tau_V = \frac{w}{V}$ is the characteristic flow time.

3. For chemical reactions

$$G(\tau) = G(0)\sum_{l-1}^{N} A_l e^{-\frac{\tau}{\tau_l}}, \qquad (16.18)$$

   where A_l is an amplitude that depends on the equilibrium concentrations and τ_s is a characteristic time that depends on the reaction rate. In the special case where there is a single first order reaction only this simplifies to a single exponential where $G(\tau) = G(0)e^{-\frac{\tau}{\tau_r}}$.

   The diagrams in Figure 16.16 show that these processes give distinct decay curves so that if the data are sufficiently good it is possible to determine which dynamic process is dominant. The difference between the diagrams is that the time is plotted on a logarithmic scale in the curves on the right.

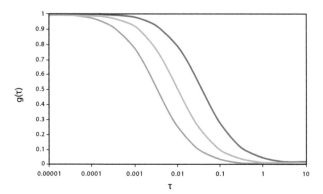

Figure 16.17: Typical autocorrelation function decay curves for diffusion for particles differing in radius by factors of three from blue to orange to red curves.

It is straightforward to solve the problem for a three-dimensional Gaussian point-spread function where

$$I(\mathbf{r}) = I_0 e^{\left(-\frac{2\left(x^2+y^2\right)}{w_0^2}\right)} e^{\frac{-2z^2}{w_z^2}} \tag{16.19}$$

and the result for the diffusion problem is

$$G(\tau) = G(0)\frac{1}{1+\frac{\tau}{\tau_D}}\frac{1}{\sqrt{1+\frac{\tau}{\kappa^2\tau_D}}}, \tag{16.20}$$

where κ is the ratio of the beam size along the z-axis and the x-y axes.

Figure 16.17 illustrates the effect on the diffusion curves when the diffusion coefficient changes be factors of three. It is fairly easy to distinguish subtle changes in diffusion coefficients. The diffusion coefficient depends inversely on the radius of a particle but as the cube root of the volume or the mass of the particle. Thus, the FCS experiment is sensitive to changes in the size, but less sensitive to the change in mass of particles.

It is interesting to note that if two or more statistically independent processes, such as diffusion and flow, apply to the same species, then the correlation functions of the two processes are multiplied.

$$G(\tau) = G(0)G_1(\tau)G_2(\tau)\ldots \tag{16.21}$$

By contrast, if the statistically independent processes apply to different molecular species observed at the same time then the correlation functions are added in proportion to the respective amplitudes for each species. An example of this would be a molecule that can exist in two states, such as free and bound, and the exchange rate is slow. In that case we expect

$$G(\tau) = G_1(0)G_1(\tau) + G_2(0)G_2(\tau) + \dots \qquad (16.22)$$

An example of this application is provided by Fujii et al. "Detection of prion protein immune complex for bovine spongiform encephalopathy diagnosis using fluorescence correlation spectroscopy and fluorescence cross-correlation spectroscopy," *Analytical Biochemistry* $370, 131 - 141(2007)$.

Note that in order to relate the amplitude of the correlation function to the number density, the correlation functions need to be normalized by the square of the average intensity as shown in Equations 13.39 and 13.40. Thus,

$$g(0) = \frac{G(0)}{\langle F(t) \rangle^2} = \frac{1}{\langle N \rangle}. \qquad (16.23)$$

This then provides a link between the autocorrelation function of the fluorescence fluctuations and the average number of molecules in the observation volume.

16.2.3 FCS of multiple species

It is possible to measure two fluorescence signals of two molecules at two different wavelengths simultaneously. In that case one would observe fluorescence fluctuations for example in the green channel, $\delta F_g(t)$ of one molecule and in the red channel, $\delta F_r(t)$ for another molecule.

The normalized autocorrelation functions for each channel would then reveal the dynamics and the number count of each species

$$g_g(\tau) = \frac{\langle \delta F_g(t)\delta F_g(t+\tau)\rangle}{\langle F_g(t)\rangle^2}$$

$$g_r(\tau) = \frac{\langle \delta F_r(t)\delta F_r(t+\tau)\rangle}{\langle F_r(t)\rangle^2}. \qquad (16.24)$$

In addition, we can calculate a cross-correlation function which interrogated whether the fluctuations in the two independent measurements are correlated. If they are, this suggests that the two chromophores are linked in their dynamics — they diffuse or flow together in space — which in turn implies that they are associated with each other on a time scale longer than the characteristic time for the movement in the confocal volume. Thus we calculate

$$g_{gr}(\tau) = \frac{\langle \delta F_g(t)\delta F_r(t+\tau)\rangle}{\langle F_g(t)\rangle \langle F_r(t)\rangle}. \qquad (16.25)$$

The decay of the cross-correlation function reflects the dynamic of the

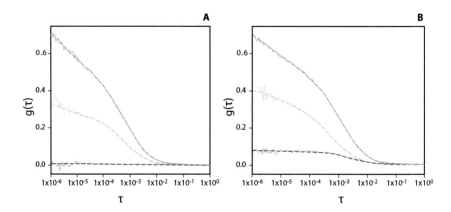

Figure 16.18: Examples of auto- and cross-correlation function decay curves before and after a period of aggregation into small aggregates.

entities that are formed by the complex of the two molecules and the decay time would reflect the size and shape of the complex.

The amplitudes of the autocorrelation functions provide an estimate of the average number of molecules of each of the individual molecules $\langle N_g \rangle$ and $\langle N_r \rangle$, and it is possible to show that the amplitude of the cross-correlation function is related proportionally to the average number of complexes, $\langle N_{gr} \rangle$, of the two molecules in the confocal volume. Thus

$$g_g(0) = \frac{1}{\langle N_g \rangle}$$

$$g_r(0) = \frac{1}{\langle N_r \rangle}$$

$$g_{rg}(0) = \frac{\langle N_{gr} \rangle}{\langle N_g \rangle \langle N_r \rangle} = \langle N_{gr} \rangle\, g_g(0) g_r(0). \qquad (16.26)$$

An example of the use of the cross-correlation function is monitoring association of molecules into small aggregates. Figure 16.18 shows the autocorrelation functions for a mixture of a protein in which some are labeled with a green fluorescent probe and others are labeled with a red fluorescent probe. At early times (the curves in a), the autocorrelation curves have the same decay times but different amplitudes reflecting that they are at different concentrations. The cross-correlation curve is zero, confirming that there are no aggregates present. At later times (the curves in b), the autocorrelation curves are nearly unchanged, but the cross-correlation curve is now visible, suggesting that there are entities in solution that contain both green and red fluorescence, supporting the notion that the proteins have aggregated to some extent. The diffusion time for the cross-correlation curve is also slightly larger, in accord with the formation of larger, slower moving aggregates.

16.3 IMAGE CORRELATION SPECTROSCOPY

The fluctuation techniques discussed so far apply to temporal fluctuations, i.e., fluctuations measured by observing the same small volume or sample as a function of time. However, the ergodic principle tells us that many average properties of a system can be measured equally well using an ensemble of samples from different regions of space. This is evident from the discussion of properties of open systems in Chapters 11 and 12. Thus rather than measuring the fluorescence emission from a confocal volume as a function of time, one can move the confocal volume to different parts of the sample and measure the fluorescence emission as a function of position. This can be very useful if the dynamic processes are very slow or absent. The net result is that the dynamic information is lost since the measurement introduces a faster dynamic — the rate of sampling across space. However, other information, such as the occupation number, is still accessible from the spatial fluctuations.

16.3.1 Image correlation spectroscopy: ICS

The implementation of spatial sampling can be achieved in a number of ways, for example moving the sample relative to the focal point of a microscope, or moving the focal point in the microscope across the sample. The latter is faster and is routine in the modern laser scanning confocal microscopes, which can produce images on time scales well below a second. An example of such an image is shown if Figure 16.19 where each of the bright spots represents a measurement of the fluorescence emission from a group of molecules located at that point in space.

> The image in Figure 16.19 was obtained by labeling the epidermal growth factor receptor on the surface of a cell in a tissue culture dish with an antibody that specifically recognizes the receptor and is marked with a fluorescent molecule. The intensity of fluorescence emission is therefore an indirect measure of the locations of the epidermal growth factor receptors on this cell.

There are several important features of this image. First, in this experiment the sample is flat so variations in fluorescence emission intensity should reflect variations in the number of fluorophores in a region. Second, the distribution of fluorescence is not uniform but is seen as a series of fluorescent spots representing a non-uniform spatial distribution of fluorophores. Third,

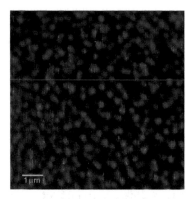

Figure 16.19: A laser scanning confocal microscope image of the fluorescence distribution from labeled proteins on a cells surface from the authors laboratory.

each of the emission regions are about the same size, which is because they represent the horizontal cross-section of the point-spread function, which is characterized by its width w. The intensity distribution across each of these spots is therefore given by the two-dimensional point-spread function in Equation 16.1. Finally, the intensity in different spots is not the same, suggesting that there is a different number of molecules in different regions of the surface. We can quantify this fluctuation in fluorescence emission through the function, $\delta F(x, y) = F(x, y) - \langle F(x, y) \rangle$, where x and y represent the coordinate in the image and $F(x, y)$ is the fluorescence measured at each point in space and $\langle F(x, y) \rangle$ is the average fluorescence intensity across the image.

We can calculate the two-dimensional autocorrelation function of the spatial fluorescence fluctuations represented by the image by calculating

$$g(\xi, \eta) = \frac{\langle \delta F(x, y) \delta F(x + \xi, y + \eta) \rangle}{\langle F(x, y) \rangle}, \tag{16.27}$$

where ξ and η are lag distances in the x and y directions (equivalent to the lag time τ in the time correlation function). Note that this is the normalized autocorrelation function.

An example of the two-dimensional autocorrelation function is shown in Figure 16.20.

Considering that the laser beam is scanning across the sample at a constant velocity, the spatial sampling is equivalent to the sample flowing under the laser beam, and hence the solution to the autocorrelation function calculation is identical to that obtained above for a flow in Equation 16.17 where $\tau_{V_x} = w/V_x = w\tau/\xi$, along the x-axis. This applies in both directions, so the autocorrelation function will be a two-dimensional Gaussian function, such that

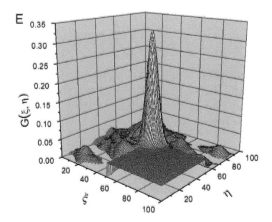

Figure 16.20: An example of the two-dimensional autocorrelation function of a confocal image with a fit to Equation 16.28 shown in the front quadrant.

$$g(\xi, \eta) = g(0,0)e^{-\left(\frac{\xi^2+\eta^2}{w^2}\right)}. \tag{16.28}$$

It is evident that this autocorrelation function contains no dynamic information. Rather the decay is determined by the width of the point-spread function, w, and therefore reflects the width of the laser beam or the diffraction limited spot, depending on the details of the microscope.

While the dynamic information is lost in this experiment, the interpretation of the amplitude of the autocorrelation function is still that same, i.e.,

$$g(0,0) = \frac{1}{\langle N \rangle}. \tag{16.29}$$

Note that this average occupation number refers to the average number of particles in the area defined by the point spread function, πw^2. We can convert this to a density by normalizing to this area such that the particle density, PD, is given by

$$PD = \frac{1}{(g(0,0)\pi w^2)} = \frac{\langle N \rangle}{\pi w^2}. \tag{16.30}$$

We can therefore obtain useful information about the density of molecules or particles on a surface by collecting an image and calculating the autocorrelation function.

It is possible that the molecules or particles on a surface are aggregated or clustered. In this case, the average occupation number refers to the average number of clusters, $\langle N_c \rangle$ and the measurement is that of the average cluster density, CD

$$CD = \frac{N_c}{\pi w^2}. \tag{16.31}$$

However, the fluorescence intensity is proportional to the number of monomeric species and hence one can estimate the average degree of aggregation, DA, by calculating the product of the average intensity and the amplitude of the autocorrelation function i.e.,

$$DA = \langle F(x,y) \rangle \, g(0,0) = c\frac{\langle N_m \rangle}{\langle N_c \rangle}, \tag{16.32}$$

where c is a constant that corresponds to the fluorescence intensity detected for a single monomer, which will depend on a number of factors (extinction coefficient, quantum yield, instrument collection efficiency, etc.), which in principle can be determined by proper calibration.

The image correlation spectroscopy approach is not limited to fluorescence microscopy but can be applied to any image in which the intensity in the image is proportional to the concentration to get an estimate of the average density of particles on a surface even if they cannot be counted individually.

16.3.2 Image cross-correlation spectroscopy: ICCS

In analogy with the fluorescence cross-correlation spectroscopy in the temporal domain, it is possible to collect images of different molecules or particles on a surface and calculate the autocorrelation functions of each as well as the cross-correlation function between the two. In this case, the density of each is determined from the autocorrelation functions and the density of particles that contain both probes is determined from the amplitude of the cross-correlation function. Thus, Equations 16.26 transform into

$$g_g(0,0) = \frac{1}{\langle N_g \rangle}$$

$$g_r(0,0) = \frac{1}{\langle N_r \rangle}$$

$$g_{rg}(0,0) = \frac{\langle N_{gr} \rangle}{\langle N_g \rangle \langle N_r \rangle} = \langle N_{gr} \rangle \, g_g(0,0) g_r(0,0). \tag{16.33}$$

Figure 16.22 shows examples of autocorrelation and cross-correlation functions for the pair of images shown in Figure 16.21.

This now provides an opportunity to quantify the extent of colocalization of different molecules on a surface. Specifically, one can estimate the fraction of all the molecules of one type that are associated with the other, thus $F(g|r)$ represents the fraction of all the green molecules that are associated with

Figure 16.21: Images collected at two wavelengths displaying the distribution of fluorescence of two molecules (A and B) and their overlapped image (C).

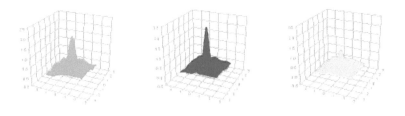

Figure 16.22: The autocorrelation functions (A and B) and the cross-correlation function (C) corresponding to the images in Figure 16.21.

the red molecules and conversely $F(r|g)$ represents the fraction of all the red molecules that are associated with the green molecules. These fractions would be calculates as

$$F(g|r) = \frac{g_{gr}(0,0)}{g_r(0,0)} = \frac{\langle N_{gr} \rangle}{\langle N_g \rangle}$$

$$F(r|g) = \frac{g_{gr}(0,0)}{g_g(0,0)} = \frac{\langle N_{gr} \rangle}{\langle N_r \rangle}. \tag{16.34}$$

Note that these fractions are generally not the same since the total number of green molecules need not be the same as the total number of red molecules.

16.3.3 Capturing dynamics

Since the image is sampling in space rather than in time, the temporal dynamics is lost in the image correlation spectroscopy approaches. However, this can be alleviated by collecting images as a function of time. Thus it is possible to calculate cross-correlation functions between images collected at different points of time as a function of the lag time between images. This then becomes an ensemble averaged FCS experiment where the dynamics of all the points in the images are captured.

From a mathematical perspective, this is captured in the spacio-temporal correlation functions given by

$$g_{ij}(\xi, \eta, \tau) = \frac{\langle \delta F_i(x,y,t)\delta F_j(x+\xi, y+\eta, t+\tau) \rangle}{\langle F_i(x,y,t) \rangle \langle F_j(x,y,t+\tau) \rangle}. \tag{16.35}$$

The cross-correlation function amplitude between the image lag time of zero and lag time τ then becomes

$$g_{ij}(0,0,\tau) = lim_{\xi \to 0, \eta \to 0} \frac{\langle \delta F_i(x,y,t)\delta F_j(x+\xi, y+\eta, t+\tau) \rangle}{\langle F_i(x,y,t) \rangle \langle F_j(x,y,t+\tau) \rangle} = \frac{1}{\langle N \rangle} f(\tau), \tag{16.36}$$

where $f(\tau)$ is a function of lag time that depends on the dynamic process that causes the particles to move on the surface during the collection of the images. This function is identical to that of the FCS experiment shown in Equations 16.16, 16.17, and 16.18 for diffusion, flow, and chemical reactions.

Figure 16.23 shows three images combined by overlaying the image collected at a lag time of zero colored in red with images collected at zero lag time and two later lag times colored in green. In the first image, all the red and green intensities are identical and the overlap is perfect and the combined color is yellow. In the second image, some of the particles have moved during the lag time and hence the green is not overlapping with the red in all locations. We therefore see red spots in the original locations, green spots in the new locations and yellow spots where the particle has not yet moved. In the third image, the lag time is long enough that almost all of the particles have moved from their original location and there are very few yellow spots left.

The cross-correlation corresponding to the first combined image is the

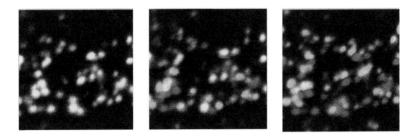

Figure 16.23: Images at zero lag time (in red) overlaid with images at three different lag times (in green).

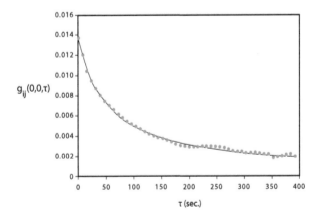

Figure 16.24: The decay of the image cross-correlation function amplitudes as a function of the lag time.

same as the autocorrelation at zero lag time and the amplitude is large. The amplitude of the cross-correlation function for the second pair of images will be smaller, and that for the third pair of images will be close to zero.

The plot of the amplitudes of the cross-correlation functions as calculated by Equation 16.36 is shown in Figure 16.24 and shows a decay that can be fit well by the equation corresponding to a diffusion process (Equation 16.16).

So the dynamics of particle movements across the surface can be determines as long as the time scale of collection of images is comparable to the time scale of the dynamic process.

CHAPTER 17

Optical properties of relevance to nanomaterials

In the previous chapters we discussed the spectroscopic properties of molecules and nanoparticles and the utility of these spectroscopic techniques. These all involve an absorption event and in some cases an emission event as well. However, all materials have other optical properties that involve interaction of the material with electromagnetic radiation that do not necessarily involve an absorption of emission. This chapter explores some of these properties with a focus on how they affect nanoscale materials.

17.1   THE WAVE EQUATION

We recall from Chapter 4 that electrons moving freely in space have both temporal and spatial components. In general, the mathematical formula for a wave — the wave equation — in one dimension is

$$\frac{\partial^2 f(x,t)}{\partial x^2} - \frac{1}{v^2}\frac{\partial^2 f(x,t)}{\partial t} = 0. \tag{17.1}$$

Here $f(x,t)$ represents the variation of the wave as a function of position and time. A specific solution to this (verify by substitution) is

$$f(x,t) = A\cos(kx - \omega t - \varphi). \tag{17.2}$$

where A is a constant reflecting the amplitude of the wave and φ is the phase of the wave.

The parameter k is the wave vector which is inversely proportional to the wavelength, λ, which describes the spatial separation of the peaks of the wave

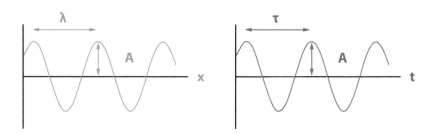

Figure 17.1: Illustration of the spatial and temporal properties of a wave.

so that $k = 2\pi/\lambda = 2\pi\kappa$. The wave number, κ, is the inverse of the wavelength: $\kappa = 1/\lambda$.

The parameter ω is the angular frequency which is inversely proportional to the period, τ, which describes the temporal separation of the peaks of the wave so that $\omega = 2\pi/\tau = 2\pi\nu$. The frequency, ν, is the inverse of the period: $\nu = 1/\tau$.

The parameter v in Equation 17.1 is the phase velocity, which is related to the wavelength and the period as $v = \frac{\omega}{k} = \frac{\lambda}{\tau}$, and describes the speed with which the peak of the wave moves either along the position axis or the time axis.

Some of these parameters are illustrated in Figure 17.1.

Note that the arguments used in Chapter 4 to define the concept of a wave packet (see Figures 4.3 and 4.4) apply equally to the general wave equations above, so that if there are uncertainties in the wavelength or period, there will be a corresponding wave packet, which in the context of the electromagnetic radiation, constitutes a photon.

17.2 ELECTROMAGNETIC RADIATION

17.2.1 The Maxwell equations

The properties of electromagnetic radiation is defined by the Maxwell Equations, which interrelate the electric field vector, $\vec{E}$, and the magnetic field vector, $\vec{B}$. They are

$$\vec{\nabla} \cdot \vec{E} = \frac{\rho}{\epsilon} \quad \vec{\nabla} \times \vec{E} = -\frac{\partial \vec{B}}{\partial t}$$

$$\vec{\nabla} \cdot \vec{B} = 0 \quad \vec{\nabla} \times \vec{B} = \mu\epsilon\frac{\partial \vec{E}}{\partial t}. \tag{17.3}$$

The Maxwell Equations depend on the properties of the medium that the electric and magnetic fields are progressing through. Thus, ρ is the charge density of the medium — in most cases there are no free charges, and hence this may be zero; ϵ is the **permittivity** of the medium, which affects both the electric and the magnetic field; and μ is the **permeability** of the medium, which affects only the magnetic field.

The divergence (the dot-product) is a scalar while the curl (the cross-product) is a vector perpendicular to the plane of the two vectors. This places the electric field in one plane and the magnetic field in a plane perpendicular to the electric field.

Recall that the Del operator, $\vec{\nabla}$, contains the partial derivatives in each of the three directions

$$\vec{\nabla} = \left(\frac{\partial}{\partial x}, \frac{\partial}{\partial y}, \frac{\partial}{\partial z} \right) = \left(\frac{\partial}{\partial x}\hat{i} + \frac{\partial}{\partial y}\hat{j} + \frac{\partial}{\partial z}\hat{k} \right). \tag{17.4}$$

The gradient is the operation of the Del operator on a scalar function and is a vector.

The divergence is the dot product of the Del operator with a vector function and becomes a scalar

$$\vec{\nabla} \cdot \vec{E} = \frac{\partial E_x}{\partial x} + \frac{\partial E_y}{\partial y} + \frac{\partial E_z}{\partial z}. \tag{17.5}$$

The curl is the cross-product of the Del operator with a vector function and remains a vector

$$\vec{\nabla} \times \vec{E} = \left(\frac{\partial E_z}{\partial y} - \frac{\partial E_y}{\partial z}, \frac{\partial E_x}{\partial z} - \frac{\partial E_z}{\partial x}, \frac{\partial E_y}{\partial x} - \frac{\partial E_x}{\partial y} \right). \tag{17.6}$$

An operation with the Del function twice is the Laplacian, which we have met before. It is a scalar, so operations on scalars produce scalars and operations on vectors yield vectors.

It is possible to show (see problem assignment) that the Maxwell Equations can be transformed into the wave equation for the electric field of the electromagnetic radiation

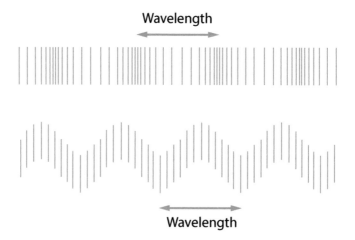

Figure 17.2: Comparison of longitudinal and transverse waves.

$$\vec{\nabla}^2 \vec{E} = \mu\epsilon \frac{\partial^2 \vec{E}}{\partial t^2}, \tag{17.7}$$

which yields the most general solution for a wave of the electric field in three-dimensional space

$$\vec{E}(\vec{r}, t) = \vec{E}_0 exp\left(i\left(\vec{k}\cdot\vec{r}\right) - \omega t\right), \tag{17.8}$$

with the wave vector amplitude related to the wavelength, the angular frequency, and the phase velocity, also known as the speed of light, $c = 1/\sqrt{\mu\epsilon}$.

The wave vector, $\vec{k}$ defines the direction of propagation of the wave

$$\left|\vec{k}\right| = \frac{2\pi}{\lambda} = \frac{\omega}{c}. \tag{17.9}$$

It is similarly possible to derive the wave equation for the magnetic field, which is identical

$$\vec{\nabla}^2 \vec{B} = \mu\epsilon \frac{\partial^2 \vec{B}}{\partial t^2}. \tag{17.10}$$

17.2.2 Light is a transverse wave

Waves can propagate either longitudinally, where the wave action is in the direction of propagation, or transversely, where the wave action is perpendicular to the direction of propagation. This difference is illustrated in Figure 17.2.

Compression waves such as pressure waves or sound waves moving through a material are examples of longitudinal waves. The waves on a surface of water are examples of transverse waves.

Consider for simplicity that our coordinate system is arranged such that the electromagnetic wave given by the Maxwell Equations propagates along the x-direction so that it is a function of x and t only. All the derivatives with respect to y and z are then zero, and if there are no charges in the medium, we find that (see problem assignments)

$$\vec{\nabla} \cdot \vec{E} = \frac{\partial E_x}{\partial x} = 0 \quad \vec{\nabla} \cdot \vec{B} = \frac{\partial B_x}{\partial x} = 0, \tag{17.11}$$

which shows that there is no variation in the component of either the electric or the magnetic field in the direction of propagation. There is therefore no longitudinal wave associated with the propagation of the electromagnetic wave. Since only the fields perpendicular to the direction of propagation can vary, the electromagnetic radiation must be a transverse wave.

Suppose that the coordinate system is further arranged such that the electric field is oriented along the y-direction, then $E_x = E_z = 0$ and hence from the Maxwell Equations we find (see problem assignments) that

$$-\frac{\partial \vec{B}}{\partial t} = \vec{\nabla} \times \vec{E} = \left(0, 0, \frac{\partial E_y}{\partial x}\right)$$

$$-\frac{\partial B_z}{\partial t} = \frac{\partial E_y}{\partial x}. \tag{17.12}$$

As a consequence, when the electric field is along the y-direction, the magnetic field is along the z-direction. These arguments are general and independent of how the coordinate system is oriented. In fact, the three vectors $\vec{k}$, $\vec{E}$, and $\vec{B}$ are always orthogonal vectors.

Using the result of Equation 17.12 we can show (see problem assignments) that

$$B_z(x, t) = \frac{1}{c} E_y(x, t), \tag{17.13}$$

which shows that the two fields are oscillating in phase and their magnitudes are different. Their units are also different, since the electric field is measured in V/m and the magnetic field is measured in $V s/m^2$ which is also known as a Tesla.

Figure 17.3 is a typical representation of an electromagnetic wave.

17.2.3 The energy in the electromagnetic radiation

The energy density, U, associated with the electric field and magnetic field, is proportional to the square of the field, so the total energy density is

$$U = U_E + U_B = \frac{1}{2}\epsilon E^2 + \frac{1}{2}\frac{1}{\mu}B^2 = \frac{1}{2}\left(\epsilon E^2 + \frac{1}{\mu}E^2\epsilon\mu\right) = \epsilon E^2, \tag{17.14}$$

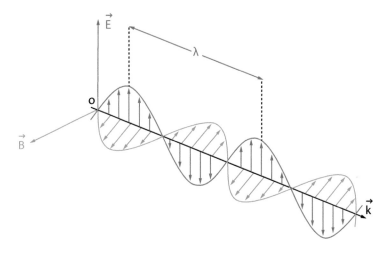

Figure 17.3: Illustration of the general electromagnetic wave.

where we have used the result that $B = E/c = E\sqrt{\epsilon\mu}$.

This shows that the energy density in the electric and magnetic fields are identical.

The energy carried by the electromagnetic radiation per unit time per unit area is represented by a vector along the direction of propagation called the Poynting Vector, $\vec{S}$, which is defined through the curl of the electric field and the magnetic field

$$\vec{S}\left(\vec{r},t\right) = \frac{1}{\mu}\vec{E}\times\vec{B} = \epsilon c^2\vec{E}\times\vec{B} = \epsilon c^2\vec{E_0}\times\vec{B_0}cos^2\left(\vec{k}\cdot\vec{r}-\omega t+\varphi\right). \quad (17.15)$$

The intensity (or irradiance), I, of the electromagnetic radiation is the magnitude of the Poynting Vector averaged over a period of the wave. This means calculating the integral of the $cos^2(\omega t)$ function which has a value of $\frac{1}{2}$ for one period. Thus

$$I = \left|\frac{1}{2}c^2\epsilon\vec{E_0}\times\vec{B_0}\right| = \frac{1}{2}c\epsilon\left|E_0^2\right|. \quad (17.16)$$

The units of the intensity is $J/m^2s = W/m^2$.

17.2.4   Photons

We saw in Chapter 4 that the quantum mechanical solution to the movement of an electron in free space led to the a wave solution for the electron, which is considered a particle. This led us to the wave-particle duality.

In a similar way, electromagnetic radiation can also be viewed as a particle, the photon, which does not have any mass. The energy of the photon is given by the frequency of the radiation, thus $U_{photon} = h\nu$, and the energy of a beam of photons is therefore $U_{beam} = Nh\nu$. Since the energy of the beam of photons must be the same as the energy of the corresponding electromagnetic radiation, U_{wave}, it is possible to calculate the flux of photons corresponding to a particular intensity.

The flux density, Ω, is interpreted as the number of photons per second per unit area and is given by the ratio of the intensity of the radiation and the energy per photon

$$\Omega = \frac{I}{h\nu}. \tag{17.17}$$

For visible light, the energy of a photon is on the order of $3 \times 10^{-19}\ J$ so for a light source, such as a laser pointer, with an intensity of $3\ mW$, we get about 10^{16} photons per second per square meter.

17.2.5 Light entering a medium

The Maxwell Equations shown above are appropriate for electromagnetic radiation in a homogeneous medium that does not interact with the electric or magnetic fields. However, all materials contain atoms with mobile electrons that can respond to the oscillating electric field. The movement of such an electron can be modeled as if it is on a spring with a particular spring constant, and hence a characteristic frequency ω_0. Using a damped forced oscillator model, the position, $x_e(t)$, of the electron can be found to be

$$x_e(t) = \left(\frac{eE_o/m_e}{(\omega_0^2 - \omega^2 - i2\omega\Gamma)}\right) e^{i\omega t}, \tag{17.18}$$

where eE_0 is the driving force, m_e, is the mass of the electron and Γ is a damping factor representing the frictional drag that the electron would experience. Thus the electron will oscillate in the medium at a frequency that depends on the applied electric field frequency, ω.

Equation 17.18 represents a complex Lorentzian function similar to what we saw in the spectroscopic absorption process in Section 14.1.4. When $\omega_0 \cong \omega$, the position can be approximated as

$$x_e(t) = \left(\frac{e}{2\omega_0 m_e}\right)\left(\frac{1}{\omega_0 - \omega - i\Gamma}\right) E(t)$$

$$= \left(\frac{e}{2\omega_0 m_e}\right)\left(\frac{\omega_0 - \omega}{(\omega_0 - \omega)^2 + \Gamma^2} + i\frac{\Gamma}{(\omega_0 - \omega)^2 + \Gamma^2}\right) E(t), \tag{17.19}$$

where $E(t)$ is the driving electric field.

The first term in the second equation is the real part and corresponds

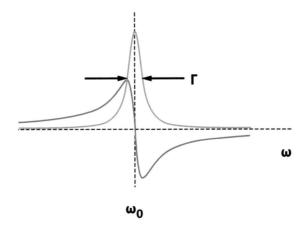

Figure 17.4: The dispersion and absorption components of the oscillating electronic components.

to the dispersion component, while the second term is the imaginary part that corresponds to the absorption component. These functions are shown in Figure 17.4.

The response of the collective of N electrons in the medium is called the induced polarization and is oriented along the direction of the electric field

$$\vec{P}(t) = Ne\vec{x}_e(t). \tag{17.20}$$

This induced polarization term can now be included in the Maxwell Equations by adding it to the time derivative of the electric field as it appears in the curl of the magnetic field such that

$$\nabla \times \vec{B} = \mu_0 \epsilon_0 \frac{\partial \vec{E}}{\partial t} + \mu_0 \frac{\partial \vec{P}}{\partial t}, \tag{17.21}$$

where the subscript 0 indicates that these are the values of permeability and permittivity in vacuum.

This can be solved to yield a new expression for the electric field in the medium as it enters it at position $x = 0$

$$E(x,t) = E_0(x = 0,t)e^{(-\alpha x/2)}e^{i(nkx-\omega t)}. \tag{17.22}$$

Here α is an absorption coefficient that causes an attenuation of the field with penetration into the material and n is the refractive index of the medium which changes the spatial wavelength but does not alter the angular frequency since the energy of the electromagnetic radiation is conserved.

The coefficients in Equation 17.22 are defined in terms of the imaginary and real parts of the coefficient of $E(t)$ in the expression of the polarization

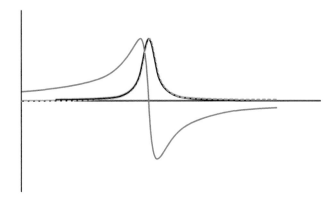

Figure 17.5: The frequency dependence of the normalized refractive index less $(n - 1$, brown) and the absorption coefficient $(\alpha$, red) in Equation 17.23, the latter compared to the absorption (black) as defined by Equation 17.19.

in Equation 17.20. Thus the absorption coefficient may be defined in terms of the absorption part while the refractive index may be defined in terms of the dispersion part of the Lorentzian function as illustrated in Figure 17.4.

$$\alpha = 2 \left(\frac{Ne^2}{4\epsilon_0 m_e \omega_0 c_0} \right) \left(\frac{\omega \Gamma}{(\omega_0 - \omega)^2 + \Gamma^2} \right)$$

$$n = 1 + \left(\frac{Ne^2}{4\epsilon_0 m_e \omega_0} \right) \left(\frac{\omega_0 - \omega}{(\omega_0 - \omega)^2 + \Gamma^2} \right) \tag{17.23}$$

The difference between these functions and those in Figure 17.4 relate to the amplitudes and to the absorption coefficient which now includes the frequency in the numerator as well. This difference is shown in Figure 17.5 and is the subject of a problem in the problem assignment.

It is evident from Equation 17.22 that the absorption coefficient α causes a decrease in the amplitude of the field as it moves further into the new medium. Since the intensity of the light is proportional to the square of the electric field, then it follows that

$$I(x) = I(0)e^{-\alpha x}, \tag{17.24}$$

which will ultimately lead to the Beer–Lambert Law for absorption of molecules in a solution (see problem assignments).

We also note that the refractive index is defined relative to the value 1, which is the refractive index of a vacuum. Because the speed of light equals the ratio of the frequency and the amplitude of the wave vector (Equation 17.9) and because the latter is modified by multiplying by the refractive index of the medium so that $k_{med} = nk$, it follows that the speed of the light in the medium, c is reduced by the factor of the refractive index as well. Thus an alternative definition of the refractive index is

$$n = \frac{c_0}{c} = \frac{\sqrt{\epsilon\mu}}{\sqrt{\epsilon_0\mu_0}} = \sqrt{\frac{\epsilon}{\epsilon_0}}, \qquad (17.25)$$

where we observe that this is a description of the electrical field propagation so that $\mu = \mu_0$.

The wavelength of the light in the medium is also reduced by the refractive index so that $\lambda = \lambda_0/n$, but the frequency, $\omega = 2\pi\nu$ of the light is unchanged since the energy is still given by $E = h\nu$.

Finally, the intensity of the light inside the medium, I_n, is increased by a factor of the refractive index relative to the intensity in a vacuum, I_0, so that $I_n = nI_0$ (see problem assignment).

The terms in the exponents of Equation 17.22 can be rearranged to give

$$E(x,t) = E_0(x = 0, t)e^{i\left[n - \frac{i\alpha}{2k_0}\right]}e^{-i\omega t}. \qquad (17.26)$$

The term in the square brackets is the **complex refractive index**

$$\tilde{n} = \left[n - \frac{i\alpha}{2k_0}\right] = \left[n - i\frac{\alpha c_0}{2\omega}\right] = n - i\kappa \qquad (17.27)$$

The two terms in the complex refractive index can be related to a complex dielectric function as

$$\tilde{\epsilon} = \epsilon_0\tilde{n}^2 = \epsilon_0\left(n^2 - \kappa^2 - in\kappa\right) \qquad (17.28)$$

from which we can get

$$n = \sqrt{\frac{1}{2\epsilon_0}\left(|\epsilon| + \epsilon_R\right)}$$

$$\kappa = \sqrt{\frac{1}{2\epsilon_0}\left(|\epsilon| - \epsilon_R\right)}, \qquad (17.29)$$

where $|\epsilon|$ is the amplitude of the dielectric constant and $\epsilon_R = n^2 - \kappa^2$ is the real part of the complex dielectric constant.

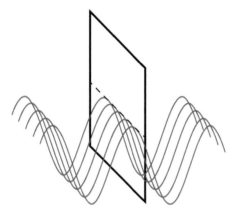

Figure 17.6: Illustration of the surface that defines a wave-front.

17.3 REFLECTION AND REFRACTION

If we consider a single source of electromagnetic radiation, the waves might all move in one direction, as in a laser beam, or might radiate spherically in all directions, as from a light bulb. In either case, it would be possible to define a surface at which the waves are in phase with all the waves at their maxima such as illustrated in Figure 17.6. This surface is called the wave-front.

There will be a wave-front at each peak, so the wave-fronts are separated in distance by one wavelength and each travel at the speed of the wave, which in turn depends on the medium as we have seen above.

Consider the general solution to the wave equation as expressed in Equation 17.8 and let the wave travel along the x-direction. In that case

$$E(x,t) = E_0 e^{i(kx - \omega t - \phi)}. \tag{17.30}$$

For the wave-front surface, the term in the exponential $(kx - \omega t - \phi)$ is a constant and it follows that x is a constant at any given time t which means that the wave-front is a surface defined by the plane perpendicular to the direction of propagation.

If the source of radiation is a point, then the wave-front will be defined by the surface of a sphere; at each point on the surface the direction of propagation is along the radii of the sphere and the wavefront is the tangential plane on the sphere at that point. This can be shown by solving the wave equation, Equation 17.7, using spherical coordinates to get

$$\vec{E}(r,t) = \frac{\vec{E_0}}{r} e^{i(kr - \omega t - \phi)}. \tag{17.31}$$

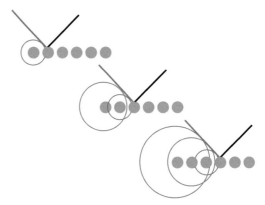

Figure 17.7: Illustration of the formation of a plane wave upon reflection.

Since the intensity is proportional to the square of the field it follows that the intensity falls off as a function of r^2 as the radiation propagates spherically. This is in contrast to the linear radiation, or the plane wave, where the intensity is constant along the direction of propagation.

Consider a wave-front of a plane wave impinging on a surface of regularly spaced objects that each diffract the light. Each object will be a source of a diffracted spherical wave as illustrated in Figure 17.7. These diffracted spherical waves will in turn interfere constructively at the edge of the spheres to form a planar wave coming off the surface at an angle that depends on the angle of incidence. This is the **reflected wave** from the surface.

In a similar fashion there will be spherical waves in the medium under the surface, but since these waves travel at a different speed in the medium, the angle of the planar wave will be different. This is the **refracted wave**.

17.3.1 Snell's Law

Figure 17.8 shows a series of wave-fronts reflected off a surface and shows that since the speed of the light is the same on the incident and reflected side of the surface, the constructive interference will occur when the reflected wave is at the same angle relative to the surface normal as the incident angle. In other words $\theta_r = \theta_i$.

Figure 17.9 shows a series of wave-fronts refracted by a surface into a medium with a larger refractive index where the speed of light is smaller and the wavelength is shorter. This means that the distance between the wave-fronts is decreased. The constructive interference will now be for the spherical wave-fronts at a smaller angle. As a consequence the angle of the transmitted wave is smaller than the incident wave.

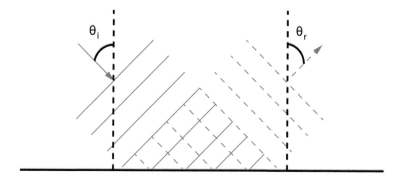

Figure 17.8: Illustration of the reflections of a series of wave-fronts.

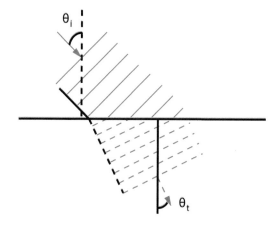

Figure 17.9: Illustration of the refraction of a series of wave-fronts.

The exact relationship between these angles is given by Snell's Law, the proof of which will be part of the problem assignment.

$$n_t sin(\theta_t) = n_i sin(\theta_i) \tag{17.32}$$

Snell's Law also predicts that if the refractive index of the medium is less than that in the incident medium, the angle of the refracted beam is larger than the incident angle.

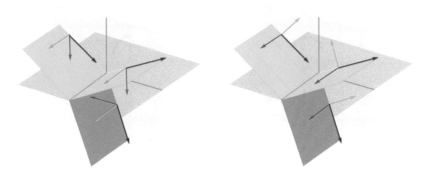

Figure 17.10: Illustration of the reflection of s-polarization (left) and p-polarization (red). The orange arrows show the direction of the electric field and the red arrows the direction of the magnetic field. The maroon arrow shows the direction of propagation.

17.3.2 The s- and p-polarization

The picture above does not address what happens to the orientation of the electric and magnetic fields as they are reflected or refracted. The polarization of the light will be important. Since the polarization is in a plane perpendicular to the direction of propagation, we need to define a series of planes relative to the reflection or refraction surface.

The plane that contains the reflection or refraction surface is termed the **plane of the interface** (because it is between two media with different refractive indexes). The plane that contains the wave vectors of both the incident $(\vec{k}_i)$ and reflected $(\vec{k}_r)$ waves is the **plane of incidence**. These two planes are necessarily perpendicular to each other since the plane of incidence contains the normal to the plane of the interface.

The plane that contains the electric field vector need not be in the plane of incidence (and neither does the plane that contains the magnetic field vector). However, the electric field vector (or the polarization) can be considered to be made up of two components, one perpendicular to the plane of incidence (the **s-polarization**) and another parallel to or in the plane of incidence (the **p-polarization**).

The fate of the electrical vector orientation upon reflection and refraction for s-polarization and p-polarization is shown in Figure 17.10

It is evident that the s-polarization component of the electrical field (orange arrows) remains in the same plane parallel to the interface and perpendicular to the plane of incidence for both the reflected and refracted beams while that of the magnetic field (red arrows) remains in the plane of incidence but changes orientation relative to the plane of the interface.

It is also evident that the p-polarization component of the electric field (orange arrows) remains in the plane of incidence, but changes orientation relative to the plane of the interface, while that of the magnetic field (red arrows) remains in the plane parallel to the interface and perpendicular to the plane of incidence.

17.3.3 Fresnel equations

The behavior of the s- and p-polarization components shown in Figure 17.10 arise from the boundary conditions used to solve the Maxwell Equations at the interface, which stipulate that the behavior of the electric and magnetic fields that are parallel to the interface must be continuous at that interface. These are the same conditions that we imposed for the solutions of the wave equations in the particle of the box at the edges of the box, whether bounded by an infinite or finite potential. The mathematical expression for this continuity is at the interface is that for s-polarization, the sum of the electrical fields in the incident and reflected beams must equal that of the transmitted beam. Correspondingly, the magnetic fields multiplied by the cosine of the angle must add to zero. Thus for s-polarization

$$E_i + E_r = E_t$$
$$-B_i cos(\theta_i) + B_r cos(\theta_r) = B_t cos(\theta_t). \tag{17.33}$$

Similarly, for p-polarization

$$B_i + B_r = B_t$$
$$E_i cos(\theta_i) + E_r cos(\theta_r) = E_t cos(\theta_t). \tag{17.34}$$

With these boundary conditions and some algebra, it is possible to derive the Fresnel Equations for perpendicular (s-polarization) and parallel (p-polarizations) light. The Fresnel Equations are defined in terms of reflection (r) and transmission (t) coefficients defined respectively as the ratio of the reflected electric field relative to the incident electric field $r = E_r/E_i$ and the transmitted electric field relative to the incident field $t = E_t/E_i$.

The Fresnel Equations for perpendicular polarized light (s-polarization) are

$$r_\perp = \frac{E_r}{E_i} = \frac{n_i cos\theta_i - n_t cos\theta_t}{n_i cos\theta_i + n_t cos\theta_t}$$
$$t_\perp = \frac{E_t}{E_i} = \frac{2n_i cos\theta_i}{n_i cos\theta_i + n_t cos\theta_t} \tag{17.35}$$

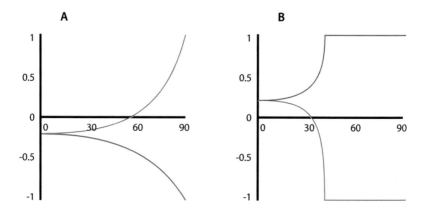

Figure 17.11: The reflection coefficients for parallel (red) and perpendicular (blue) polarization at the interfaces between air and glass (A) air to glass and (B) glass to air as a function of the angle of incidence.

and those for parallel polarized light (p-polarization) are

$$r_{\parallel} = \frac{E_r}{E_i} = \frac{n_i cos\theta_t - n_t cos\theta_i}{n_i cos\theta_t + n_t cos\theta_i}$$
$$t_{\parallel} = \frac{E_t}{E_i} = \frac{2n_i cos\theta_i}{n_i cos\theta_t + n_t cos\theta_i}. \qquad (17.36)$$

The Fresnel Equations are useful since they tell us how effectively light is reflected off a surface as a function of the angle of incidence. The reflection coefficients for the parallel and perpendicular components of the incident light are shown in Figure 17.11 as a function of the incident angle and for two cases: (A) when the refractive index changes from 1 in air to 1.5 in glass at the interface and (B) when the refractive index changes from 1.5 in glass to 1 in air at the interface, i.e., when the light enters glass and when it leaves the glass.

There are several important insights to get from these graphs.

First, there is no difference in the reflection of the two polarizations as the angle of incidence approaches zero, that is, as the light arrives perpendicular to the interface (recall that the angle of incidence is measured relative to the normal of the interface) and for both polarizations there is complete reflection as the angle of incidence approaches 90° (both positive and negative values of the reflection coefficients correspond to real reflection).

Second, there is an angle at which the parallel polarization is not reflected

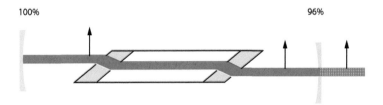

Figure 17.12: Schematic of a typical laser.

at all ($r_{||} = 0$). This angle is called the **Brewster's Angle** and depends on the arctangent of the ratio of the refractive index of the two materials.

$$\theta_B = arctan(n_t/n_i) = tan^{-1}(n_t/n_i). \tag{17.37}$$

For this case with air and glass, the Brewster's angle is 56.3°, as measured from the air side of the interface, both at entry and at exit. Note that the Brewster's angle appears to be lower in Figure 17.11B since in this plot the angle of incidence is in the glass medium.

Third, as the light is reflected from the glass to the air, the angle at which total reflection occurs is much less than 90°. This shows that above this angle, called the **critical angle**, there is always complete reflection. This is also termed total internal reflection. Total internal reflection allows light to be confined to a medium of higher refractive index, such as an optical fiber. Note that the critical angle is the same for the two polarizations and it can be calculated from the arcsin of the ratio of the refractive indexes

$$\theta_c = arcsin(n_t/n_i) = sin^{-1}(n_t/n_i). \tag{17.38}$$

For the glass-air interface, the critical angle is about 41.8° as measured on the glass side.

The laser relies on the Brewster's Angle   The effect of complete transmission at the Brewster's angle is used in lasers to discriminate between the two polarizations of light and therefore generate a coherent and vertically polarized laser beam. Figure 17.12 shows schematically how a typical laser is constructed.

The region in the middle is designed to create a plasma of molecules or

Figure 17.13: Schematic of an optical fiber transmitting light through total internal reflection.

atoms in the excited state, usually by electrical excitation. The emission is reflected back into the plasma by the mirrors at either side. As the emission returns, it will induce stimulated emission from the excited molecules in the plasma. This processes continues until the rate of stimulated emission is in a steady state with the rate of excitation. The distance between the mirrors create a cavity whose length determines the exact wavelength that will provide a constructive interference within the laser. This represents the *light amplification by stimulated emission of radiation* — the **laser**. As the light passes through the end prisms that are constructed to be at the Brewster's angle, there is a selective amplification of the parallel polarization over the perpendicular polarization resulting in a polarized laser beam. The mirror on the right in the diagram is created to allow a small (typically about 4%) amount of light to be transmitted allowing for the laser light to emerge. The slight curvature of the end mirrors cause the focusing of the light and as a consequence there is a slight divergence of the laser beam as it exits.

Optical fibers rely on the critical angle   The total internal reflection arising at the interface between high and low refractive index materials allows us to construct fibers that will retain almost all of the light with little loss along the length of the fiber. This is illustrated in Figure 17.13.

17.3.4   Reflectance and transmittance

The reflection and transmission coefficients do not directly tell us how much light is reflected or transmitted, so we define the reflectance and transmittance in terms of the relative power being reflected or transmitted, i.e.,

$$R = \frac{I_r A_r}{I_i A_i} \frac{n_i \epsilon_0 c_0 / 2\, |E_0|^2\, A_r}{n_i \epsilon_0 c_0 / 2\, |E_0|^2\, A_i} = r^2$$

$$T = \frac{I_t A_t}{I_i A_i} \frac{n_t \epsilon_0 c_0 / 2\, |E_0|^2\, A_t}{n_i \epsilon_0 c_0 / 2\, |E_0|^2\, A_i} = \frac{n_t}{n_i} \frac{cos(\theta_t)}{cos(\theta_i)} t^2, \tag{17.39}$$

where $R + T = 1$.

17.3.5 Evanescent waves

In general terms we can rewrite the general solution to the wave equation in Equation 17.8 as

$$\vec{E}(\vec{r}, t) = \vec{E}_0 e^{i(k_x x + k_y y)} e^{i k_z z} e^{-iwt}, \tag{17.40}$$

where we have used the fact that real values for the wave vector components along each dimension (k_x, k_y, k_z) leads to a propagating wave through space, this constitutes the plane wave that we have discussed so far.

Constraining all the wave vector components to be real is not necessary, so consider what happens if one of the wave vector components is imaginary, say $k_z = ik_z'$, where k_z' is real. In that case, we get

$$\vec{E}(\vec{r}, t) = \vec{E}_0 e^{i(k_x x + k_y y)} e^{k_z' z} e^{-iwt}, \tag{17.41}$$

which represents a wave that decays exponentially along the z-direction. Correspondingly if two components were imaginary, we would have a wave propagating in one direction and decaying in the other two directions.

Equation 17.41 is the mathematical representation of an evanescent wave, which contrasts with the plane wave by having a decay along one direction perpendicular to the direction of propagation.

Recall from the Fresnel equations that above the critical angle the reflection of the light is total, but at the interface there must be an interaction between the electromagnetic radiation and the medium at the other side of the interface. In fact, there is a requirement for continuity of the electric field at this interface. This suggests that there must be a component of the electric field in the lower refractive index medium as well.

If the reflecting interface is set to be in the x, y-plane, then there will be an evanescent wave at the interface as described by Equation 17.41. This wave oscillates in the x, y-plane with a wavelength given by k_x and k_y and a frequency given by ω and in addition, the electric field decays exponentially perpendicular to the interface into the low refractive index medium as shown in Figure 17.14.

While there is no power in the evanescent wave since the reflected power is equal to the incident power, the electric field along the z-direction can still interact with molecules in the low refractive index medium causing absorption or emission of light.

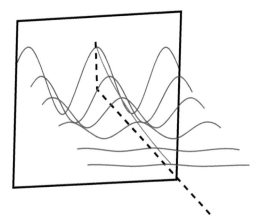

Figure 17.14: Illustration of an evanescent wave decaying along the direction perpendicular to the surface.

It is interesting to note that mathematically the phenomenon of a decaying electric field of an evanescent wave in the low refractive index medium is equivalent to the decay of the wave functions for electrons into the region of space where the potential is greater than the energy of the electron (See Section 4.3.1).

If the wave is confined to a one-dimensional optical fiber, the evanescent wave will decay radially from the surface which can be expressed by an equation analogous to Equation 17.41 with two of the wave vector components being imaginary (see problem assignment)

The phenomenon of total internal reflection and the corresponding evanescent wave is exploited in a particular type of fluorescence microscopy — total internal reflection fluorescence (TIRF) microscopy. Here the excitation light from a laser is reflected off the internal surface of a high refractive medium (quartz) to create an evanescent wave in the medium (water) immediately above it. Molecules close to the surface, typically within the first 100 nm or so, will be excited and their emission can be captured with a microscope objective. In this manner, molecules close to a surface can be studied selectively. This has been useful in studies of adhesion properties of cells to surfaces and rates of absorption of molecules to a surface. In some sense this is another form of spatially selective microscopy akin to the confocal and superresolution microscopy tools discussed in Chapter 16.

17.4 SURFACE PLASMONS

17.4.1 Dielectric behavior of metals

In an earlier chapter (Section 5.1.2) we introduced the concept of electrons in a metal behaving like a gas — the free electron or Fermi gas model. These electrons in a metal should be able to interact with the oscillating electric field of the electromagnetic radiation.

It is possible to show that for metals the complex dielectric function is a function of frequency

$$\epsilon(\omega) = \epsilon_0 \left(1 - \frac{\omega_p^2}{\omega^2 + i2\omega\Gamma} \right), \tag{17.42}$$

where $\omega_p = \sqrt{Ne^2/\epsilon_0 m_e}$ is called the plasma frequency (the frequency with which the electrons would oscillate when subjected to a small force as if they were on small springs) and Γ is a damping function.

For most metals, the plasma frequency is several orders of magnitude larger than Γ, so $\omega \gg \Gamma$ and we get that $(\epsilon(\omega) = \epsilon_0 (1 - \omega_p^2/\omega^2))$, which is a real number that is positive above the plasma frequency and negative below the plasma frequency.

When the dielectric constant is negative below the plasma frequency, the refractive index is imaginary and hence the electromagnetic radiation decays rapidly within the metal. On the other hand, when the dielectric constant is positive above the plasma frequency the refractive index is real and the metal can transmit the electromagnetic radiation. Thus metals are highly reflective at low frequencies (long wavelengths such as visible light) and highly transparent at high frequencies (short wavelengths such as x-rays).

17.4.1.1 Interface waves

Let us consider the possibility of an electromagnetic wave traveling in the plane of an interface between two media. We can orient the interface in the x, y-plane and the direction of propagation of the wave along the x-direction. We expect that the electric fields are along the x-direction and the z-direction and the magnetic field direction is along the y-direction. In that case the electric fields and magnetic fields are described, for each medium, by

$$\vec{E}_l = (E_{lx}, 0, E_{lz})e^{ik_x x - \kappa_l |z| - i\omega t}$$
$$\vec{B}_l = (0, B_{ly}, 0)e^{ik_x x - \kappa_l |z| - i\omega t}, \tag{17.43}$$

where $l = 1, 2$ denoting each medium.

When these are introduced into the Maxwell Equations and solved for the boundary conditions relevant to the surface, the result is that the in order for these surface waves to exist, we must satisfy the condition

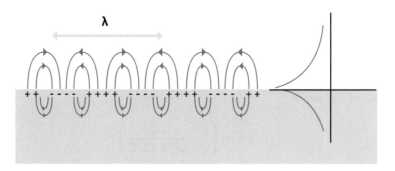

Figure 17.15: Illustration of the surface electron oscillations leading to oscillating electric fields in the media.

$$\frac{\epsilon_1}{\kappa_1} = -\frac{\epsilon_2}{\kappa_2} \qquad (17.44)$$

and since the κ values are always positive, the dielectric constants must have opposite signs.

The dielectric constants for metals are negative for frequencies below the plasma wave frequency, so a metal in contact with a normal dielectric medium can support such a wave.

The collective oscillations of electrons at the surface in the presence of electromagnetic radiation constitute a surface plasmon polariton. By Equation 17.43 it follows that the surface plasmon polaritons decay exponentially into the media on either side of the interface (since k_l is real and positive) and that they oscillate in the plane of the interface with a spatial wavelength, λ, determined by k_x and a frequency determined by ω. The concept of the surface plasmon polariton is shown in Figure 17.15 where the electron oscillations cause charge separations with field lines that couple into the media to different extents since κ depends on the refractive index. The decay distance is shorter in the higher index medium.

In the bulk of the electron gas model there are similar electron oscillations that create bulk plasmons, which in turn determine the plasma frequency ω_p. The energy of these bulk plasmons is then $E_p = \hbar\omega_p$.

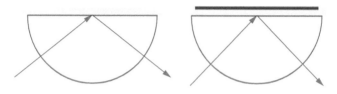

Figure 17.16: Illustration the configuration of surface plasmon resonance measurements.

17.4.2 Surface plasmon resonance

The surface plasmons will interact with external electromagnetic fields and will couple and lead to absorption of the added light. This causes the phenomenon of surface plasmon resonance. In order for this coupling to occur, the momentum of the incident photons must match that of the surface plasmons. This can be achieved by illuminating through a block of glass (with lower refractive index that the metal) where the change in angle will alter the momentum of the p-polarized component of the light. At the appropriate combination of wavelength and angle, the momentum match causes an absorption of light. Since the coupling is sensitive to the refractive index match at the interfaces, the angle at which absorption of light is maximal will depend on whether there is other material adsorbed on the surface. This then allows for a sensitive detection of adsorption of biological materials such as proteins onto the surface. The configuration used in a typical surface plasmon resonance experiment uses a thin film of metal where the surface plasmon is sensitive to the refractive index match on both sides. This is illustrated in Figure 17.16. As material adsorbs to the surface, the angle at which the maximal absorption occurs will change.

17.4.3 Plasmon resonance in nanoparticles

When nanoparticles are made of metals or coated with metal films, the electrons in the nanoparticles will couple with an external electromagnetic field to create a localized plasmon. This cannot propagate since the nanoparticle is much smaller than the wavelength of the light, however they will have a characteristic resonance frequency that causes absorption of the light at particular frequencies. This is plasmon absorbance which is responsible for giving nanoparticles of materials their color. The plasmon resonance in the nanoparticles will depend on their size and hence the peak of the plasmon absorption is a reliable measure of the average size of the nanoparticle.

Equally interesting, if two or more nanoparticles come in close contact, their plasmons can couple with each other because of the exponential decay of the electric field away from their respective surfaces. This coupling in turn leads to a change in the absorption, generally a shift to longer wave lengths (red-shift) and a broader absorption line. This effect has been used extensively as the basis for optical sensing devices.

These sensors are designed such that the metallic nanoparticles (typically made of gold since it is fairly easy to chemically modify the surface using sulfhydryl chemistry) are coated with molecules (for example antibodies) that will recognize the analyte of interest. When the analyte is present, it may bind to multiple nanoparticles, causing these to aggregate and change the absorption properties of the solution.

CHAPTER 18

Epilog and acknowledgments

18.1 FINAL THOUGHTS

The objective of this book has been to establish some of the foundational knowledge that helps us understand some of the phenomena that arise in small systems. The focus has been on this foundational knowledge since it is enduring. While we may gain further and deeper insight in to the special world of nanoscience, the principles reiterated in this text are not likely to change in the near future.

Much of what has been presented in this book has been known for many years and is often taught in the undergraduate curriculum in chemistry and physics departments in universities and colleges around the world. However, the fields of nanoscience and nanotechnology increasingly attract and need contributions from individuals with less formal training in the relevant topics. This book has therefore been designed to fill a gap for those who want to understand why the world at the nanoscale is different in fundamental ways.

At first sight, the chapters cover a range of unrelated topics, but one of the possibly unique aspects of this text is that it presents in one place rather diverse topics that are all in some way relevant to understand the nature of the nanoscale world or how we study it.

While it is tempting to carry on with more detailed discussions of the world of nanoscience and nanotechnology, it was a deliberate choice not to do so for a couple of reasons. First, the fields of applications move very rapidly and any book that attempts to describe the current state will soon be out of date. Second, the applications are endless and it is not clear where one would focus to get the most impact. Third, while nanoscience and nanotechnology applications will continue to evolve, they will in the future not necessarily be known by those terms, but rather their fields of applications, whether that be sensors in geomatics, diagnostic tools in medicine, or catalysts for the energy field. I am therefore hoping to focus on the more enduring foundational knowledge for those who need it, but were never exposed to it.

Clearly there are deficiencies in the structure of this book which arise from my limited knowledge of some topics. For example, this text does not address the extremely important topics of magnetics or mechanics at the nanoscale. Perhaps future editions will address such subjects. Also, some topics are possibly covered in more detail than some would wish, but that is motivated by personal interest in topics such as fluctuations in systems at low concentration.

It is not the intent of this book that it be used as a single reference in any given course, nor that all the topics should be part of any given curriculum. Hopefully, however, there are elements of this text that can prove useful in some courses and more importantly as a bridge towards more advanced exploration of the key topics either in advanced courses or in research environments.

Naturally, if this text enjoys any measure of success, there may be additional editions. Any suggestions for improvement, including corrections of errors and expansion or contractions of topics, will therefore be very welcome, both from students and teachers who may be using this text.

18.2 ACKNOWLEDGMENTS

This book would not have been possible without support from family, mentors, colleagues, past students, and friends.

My wife, Pat, has been a tower of strength through our joint journey from London (Ontario) to Edmonton via Pasadena, Ithaca, St. Louis and London again. During this voyage, she has been supportive, encouraging, and most importantly, willing to face the risks that were associated with each move as my career went from student to post-doctoral fellow, to professor, to department chair, vice-president of research and director general and back to professor again. It has been a great experience made better by her company at every step. We are also blessed with two great children, Lars and Anne, who have embarked with great success on their own journeys with their partners and children. I am also grateful to my parents for providing the opportunity to immigrate to Canada and for their support as I stayed when they moved back to Denmark.

I have had many mentors who have guided, directed, and contributed to my success and I will mention a few who had particular impact on my choice of career steps as they evolved: Peter Guthrie, Albert Stoessel, and Jake Stothers had faith in my ability and guided me through undergraduate research experiences that had profound impact on my interest in performing experiments. Duncan Hunter, Garth Kidd, and Colin Baird were instrumental in guiding me to pursue opportunities in the US which had a significant influence on my outlook on education and research. Sunney Chan, my doctoral supervisor at Caltech, challenged me, supported me, and gave me the freedom to explore a number of directions in my graduate research. This experience taught me to look at the big picture and to understand the value of interdisciplinary research and the need to draw on multiple fields of expertise to get the tools to attack complex problems in the field of membrane biophysical chemistry. Elliot Elson, my post-doctoral supervisor at Cornell and Washington University Medical School, exposed me to a whole new world of biophysics and cell biology, and with Watt Webb, allowed me to pursue some of my own interests that ultimately provided a platform for my independent research programs at Western.

As I was learning to teach subjects as diverse as thermodynamics, quantum mechanics, polymer chemistry, spectroscopy, environmental chemistry, and membrane biochemistry I received invaluable support from all my colleagues in the Departments of Chemistry and Biochemistry at the University of Western Ontario, particularly Bill Meath, Patrick Jacobs, Colin Baird, Rob Lipson, Derek Leaist, Duncan Hunter, and Ted Lo.

In terms of writing this book, I owe the most to my former students — undergraduate and graduate — who patiently listened to my lectures and my explanations and provided the feedback that led to improvements in my own teaching and mentoring and ultimately laid the foundation for my own understanding of the material that forms the core of this text. I also wish to thank the Department of Chemistry at University of Alberta for the opportunity and freedom to return to the department to teach some specialty courses whose content formed the foundation for the scope and the structure of the text. The students in those courses affirmed my belief that this text might be useful for individuals who want to enter the complex fields of nanoscience and nanotechnology from disciplines where the grounding in the key physical chemical principles are limited.

I have too many friends to thank them individually, but you know that I appreciate the friendship and support that you have provided which has allowed me to stay grounded and at the same time enjoy life in all its aspects.

VI

Problems and references

CHAPTER 19

Problems

There are different types of problems: those that are intended to remind the reader about the symbolism and the structure of equations; those that require comprehension of the principles and the concepts; and those that are intended for the advanced reader who wants to understand the details of derivations or calculations. These latter problems are designated by an asterisk (*).

19.1 CHAPTER 1 PROBLEMS

Problem 1.1 What are some common features of the various definitions of of nanomaterials?

Problem 1.2 Consult the web-site for the Emerging Nanotechnologies project (www.nanotechproject.org) and discuss some of the most frequent applications of nanotechnology in consumer products and speculate on why these are so common.

19.2 CHAPTER 2 PROBLEMS

Problem 2.1 Without consulting Table 2.1, define each of the terms in the Time-Dependent Schrödinger Equation

$$\hat{H}\Psi(x,y,z;t) = \left[-\frac{\hbar^2}{2m}\nabla^2 + V(x,y,z) \right]\Psi(x,y,z;t) = -\frac{\hbar}{i}\frac{\partial}{\partial t}\Psi(x,y,z;t).$$

$$(19.1)$$

Problem 2.2 Using the operators in Table 2.1, calculate the result of the following operations:

1. $\left[\hat{x}\left(ax^2 + b\right) \right]$

2. $\left[\hat{p}\left(ax^2 + b\right) \right]$

3. $[\hat{x}\,(e^{-ax})]$

4. $[\hat{p}\,(e^{-ax})]$

5. $\left[\hat{p}^2\,(cos(ax))\right]$

Problem 2.3 Use any test function, such as one of those in Problem 2.2, to show that the position operator and the momentum operator do not commute.

19.3 CHAPTER 3 PROBLEMS

Problem 3.1 Show that for conservative forces, the wave function in Equation 3.3 satisfies the TDSE leading to Equation 3.5, the TISE.

*Problem 3.2 Show that for conservative forces, Equation 3.1 can be separated into two differential equations (Equation 3.2 and 3.5) where the constant is the energy, E.

19.4 CHAPTER 4 PROBLEMS

Problem 4.1 Show that $\psi(x)$ in Equation 4.3 satisfies the TISE in Equation 4.2.

Problem 4.2 Show that $\psi(x)$ in Equation 4.3 is NOT an eigenfunction of the momentum operator.

Problem 4.3 Solve the TISE for $V(x) = V_0$ for all values of x to show that the momentum for a particle moving along the positive x-axis is $p = \sqrt{2m_p(E - V_0)}$ with $E > V_0$.

*Problem 4.4 Use a spreadsheet or computer programming language to reproduce Figure 4.3.

*Problem 4.5 Use the program in Problem 4.4 to show the effect of changing the uncertainty in k values.

*Problem 4.6 Show that Equation 4.21 follows from Equations 4.17 and 4.20.

Problem 4.7 Use Equations 4.21, 4.22 and 4.23 to show that $A = 1/(i\sqrt{2b})$.

Problem 4.8 Show that substituting $\psi(x)$ given by Equation 4.24 into the TISE yields the energy given in Equation 4.25.

Problem 4.9 Show that $\Delta E_n = (2n+1)h^2/(8m_p b^2)$.

Problem 4.10 Solve Equation 4.26 using Equation 4.27 to get Equations 4.28 and 4.29.

*Problem 4.11** Show that when the wave functions multiply the energies add.

Problem 4.12 Show that $E_{3,4} = E_{4,3}$ for a two-dimensional square box with infinite walls.

Problem 4.13 Describe the concept of degeneracy and give an example for a 2-D box with infinite walls.

Problem 4.14 What is the maximum number of degenerate states in a 3-D cubic box?

*Problem 4.15** Show in detail that for

1. cylindrical coordinates,

$$\frac{\partial}{\partial x} = \cos\phi \frac{\partial}{\partial r} - \frac{\sin\phi}{r}\frac{\partial}{\partial\phi}; \tag{19.2}$$

2. spherical coordinates,

$$\frac{\partial}{\partial x} = \sin\theta\cos\phi\frac{\partial}{\partial r} + \frac{\cos\theta\cos\phi}{r}\frac{\partial}{\partial\theta} - \frac{\sin\phi}{r\sin\theta}\frac{\partial}{\partial\phi}. \tag{19.3}$$

*Problem 4.16** Show in detail that for spherical coordinates

$$\nabla^2 = \left(\frac{1}{r^2}\frac{\partial}{\partial r}\left(r^2\frac{\partial}{\partial r}\right) + \frac{1}{r^2\sin\theta}\frac{\partial}{\partial\theta}\left(\sin\theta\frac{\partial}{\partial\theta}\right) + \frac{1}{r^2\sin\theta}\frac{\partial^2}{\partial\phi^2}\right). \tag{19.4}$$

Problem 4.17 Normalize $\psi(r_0, \phi, 0)$ in Equation 4.47.

Problem 4.18 Show that the energy for a particle on a ring bounded by an infinite potential is given by Equation 4.49.

Problem 4.19 Show that $E_{ring} = E_{box}$ when $b = 2\pi r_0$, whenever n=2m.

Problem 4.20 . Show that the probability distribution for the spherical harmonic Y_1^1 is independent of the angle ϕ.

Problem 4.21 Show that the energy in Equation 4.68 results from Equation 4.66 when $r = r_0$ and $V(r_0) = 0$.

Problem 4.22 Confirm Equation 4.69 using Equation 4.68.

Problem 4.23 Use a spreadsheet or computer program to calculate $R_{1,0}(r)$ and $R_{2,1}(r)$ with $Z = 1$ and $d_0 = 1$ and plot these as a function of r, in units of a_0, from 0 to $3a_0$.

Problem 4.24 For the Hydrogen Atom,

1. Evaluate E_1, E_2, E_3, and E_4 and plot these.

2. Calculate $\Delta E_{1,2}$, $\Delta E_{1,3}$, and $\Delta E_{1,4}$.

3. Calculate $\Delta E_{1,\infty}$, $\Delta E_{2,\infty}$, and $\Delta E_{3,\infty}$ [this corresponds to the H-lines in the emission spectrum of the H-atom].

Problem 4.25 Use a spreadsheet program or a computer program to calculate and plot the Hermite polynomials and the wave functions for the simple harmonic oscillator i.e.,

1. $H_0(q)$, $H_1(q)$, and $H_2(q)$

2. $\psi_0(q)$, $\psi_1(q)$, and $\psi_2(q)$.

Problem 4.26 Show that $\Delta E_n = \hbar\omega$ for the simple harmonic oscillator.

Problem 4.27 Calculate the energy differences for the rigid rotator: ΔE_1, ΔE_2, ΔE_3, and ΔE_4.

19.5 CHAPTER 5 PROBLEMS

Problem 5.1 Derive Equation 5.3.

Problem 5.2 Find the structure of beta-carotene, determine the number of conjugated double bonds, and estimate the length of the box and the number of electrons in them. Then use the particle in the box to estimate the wavelength of light expected to be absorbed by this molecule.

Problem 5.3 Find the maximum wavelength for the absorption spectrum of trans-retinal and estimate, using Equation 5.3, the effective size of the box occupied by the electrons and compare this to the estimated length of the conjugated system in the molecule.

Problem 5.4   What are the assumptions of the Fermi gas model?

Problem 5.5   Using the Fermi–Dirac statistics, calculate the fraction of electrons at the energy of the chemical potential.

*Problem 5.6   Using the properties of a wave function defined by Postulate I, set up the boundary conditions of the periodic potential illustrated in Figure 5.5 and write out the four equations that need to be solved to get to Equation 5.14. You do not need to solve them.

Problem 5.7   Describe qualitatively, the difference between a semiconductor and an insulator.

Problem 5.8   Describe qualitatively, the difference between a direct and an indirect band gap semiconductor and explain why it is important.

Problem 5.9   Describe qualitatively, the difference between an n-doped and a p-doped semiconductor.

Problem 5.10   Describe zero-, one-, two-, and three-dimensional structures and give examples of each.

Problem 5.11   Provide a rationalization for the trends observed in Figure 5.16.

Problem 5.12   Using the data in Figure 5.22, discuss whether there is a difference between a small cluster of molecules and a small nanoparticle of the same material.

19.6   CHAPTER 6 PROBLEMS

Problem 6.1   Describe the differences between isolated, closed, and open systems and give an example of each.

Problem 6.2   What are the key characteristics of isochoric and adiabatic processes?

Problem 6.3   State the first, second, and third laws of thermodynamics.

Problem 6.4   Define the heat capacities at constant volume and at constant pressure in terms of thermodynamic functions.

*Problem 6.5   Show that the equalities in Equations 6.34 and 6.35 follow from the definitions of the Helmholtz and Gibbs free energy, respectively, and expressions for the Clausius Inequality.

Problem 6.6   Provide four different criteria for spontaneous change at different conditions.

Problem 6.7   Show that equations 6.36 to 6.39 are correct.

*Problem 6.8   Given that the property of an exact differential

$$df(x, y) = \frac{\partial f}{\partial x}dx + \frac{\partial f}{\partial y}dy \qquad (19.5)$$

requires that

$$\frac{\partial^2 f}{\partial x \partial y} = \frac{\partial^2 f}{\partial y \partial x}. \qquad (19.6)$$

Derive Equations 6.40 to 6.43 from Equations 6.36 to 6.39.

Problem 6.9   Define the chemical potential in terms of the Helmholtz free energy.

19.7   CHAPTER 7 PROBLEMS

Problem 7.1   Explain qualitatively why the chemical potential of a gas decreases more rapidly with temperature than the chemical potential of a solid and why the chemical potential of a gas increases more rapidly with pressure than the chemical potential of a liquid.

Problem 7.2   Draw schematically the chemical potential as a function of temperature at a pressure corresponding to the triple point.

Problem 7.3   Explain why water does not freeze at the high pressures at the bottom of a deep lake.

*Problem 7.4   Derive Equation 7.4 from 7.2 assuming ideal gas behavior.

Problem 7.5   Consider a hollow tube with a bubble at each end with the bubble at the left being bigger than that at the right and with a continuous connection of air between them through the tube. This is an unstable situation, so what will happen next?

*Problem 7.6   Derive the Kelvin Equation 7.10 from the Laplace Equation 7.8 and Equation 7.9.

Problem 7.7   Explain the observation in Figure 7.7 that the vapor pressure of mercury increases more rapidly than water as the size of the drop decreases.

Problem 7.8   Calculate the critical radius for a water droplet at room temperature when the pressure is 10% above the vapor pressure.

Problem 7.9   Use the vector diagram in Figure 7.11 to derive Young's equation (Equation 7.18).

Problem 7.10   Show that Equation 7.19 follows directly from the balance of pressures as shown in Figure 7.13.

Problem 7.11   The xylem in trees are small tubes created from dead cells and can have diameters as small as about 25 microns. Calculate the theoretical height of the water column that such a tube could support. You should find that this is not likely enough to support the water rising to the top of a tall tree. Explore alternative explanations for the rise of water to the top of a very tall tree.

*Problem 7.12   Derive Equation 7.43.

Problem 7.13   Provide a possible rationale for the difference in the variation of binding energy with the size of spherical particles of gold (Au) and silicon (Si) illustrated in Figures 7.16 and 7.18.

Problem 7.14   Consider the following chemical reactions. For each, write an expression for the reaction coefficient, Q.

1. $CO_2(g) + H_2O(g) \rightleftharpoons H_2CO_3(s)$
2. $2CO(g) + O_2(g) \rightleftharpoons CO_2(g)$
3. $N_2(g) + 3H_2(g) \rightleftharpoons 2NH_3(g)$
4. $CuO(s) + 2NaHSO_4(aq) \rightleftharpoons Na_2SO_4(aq) + H_2O(l)$

19.8   CHAPTER 8 PROBLEMS

Problem 8.1   Explain how the density of states in the valence band is observed in the photoelectron spectrum.

Problem 8.2 What is the origin of inelastically scattered electrons in the photoelectron spectrum?

19.9 CHAPTER 9 PROBLEMS

Problem 9.1 Describe some important attractive forces between particles.

Problem 9.2 Describe some important repulsive forces between particles.

Problem 9.3 Using the expressions for the Hamaker constant in Equations 9.6 and 9.7, calculate its value for a system of two drops of water in air. Further, calculate the attractive force between them assuming they both have a radius of 100 nm.

Problem 9.4 Calculate the Debye length

1. in a 60-mM solution of $NaCl$.

2. in a 20-mM solution of $CaCl_2$.

3. in a 10-mM solution of $AlCl_3$.

Problem 9.5 Describe the concept of the diffuse double layer.

Problem 9.6 How does the zeta-potential relate to the Stern Layer?

Problem 9.7 Use a spreadsheet to calculate the overall interaction energy between two 5-nm-diameter gold spheres in an aqueous solution with 10 mM of $NaCl$. Compare this to the interaction energy between one of these spheres and a flat surface of gold.

Problem 9.8 Show for the attractive potential in Equation 9.34 reduced to that of the DLVO theory when $r_r \gg D$.

Problem 9.9 Describe the conditions that may lead to a dendritic aggregate rather than a compact aggregate.

19.10 CHAPTER 10 PROBLEMS

Problem 10.1 Explain the differences between the mean, the median, and the mode.

Problem 10.2   Calculate and plot on the same graph the binomial distribution for $n = 20$ and $p = 0.5$; the Poisson distribution with $\mu = 10$, and the normal distribution with $\mu = 10$ and $\sigma = \sqrt{10}$. Compare these in the context of the law of rare events and the central limit theorem.

Problem 10.3   Assuming x can take on the values $[7, 7, 8, 8, 8, 9, 10]$, calculate the inverse of the mean of x, $1/\langle x \rangle$, as well as the mean of the inverse of x, $\langle 1/x \rangle$ and compare these numbers.

Problem 10.4   Consider that you want to determine the speed of the water in a stream. You decide to use the "Poohsticks Game (from Winnie the Pooh books)" in which you infer the speed of the stream from the speed of an object moving on top of the stream of water. In other words, you measure the time that the object moves a fixed distance (from one side of a bridge to the other). You repeat this experiment multiple times to get a set of measurements of time. You further want to do some statistical analysis of these data. a) When is it appropriate to average the time measurements and then divide this average into the known, fixed distance to get an average speed? b) When is it appropriate to calculate the individual speeds for each time measurement and then average these calculated speeds? (Similar questions arise in many experiments where the parameter of interest is inversely related to the parameter that can conveniently be measured. An example is the classical physics experiment in which we measure the viscosity of a fluid by measuring the time a small ball of radius, r, takes to descend a fixed distance in a column of the liquid. The viscosity, η, is then calculated from Stoke's Law $F_{gravity} = 6\pi \eta r v$, where v is the calculated velocity).

19.11   CHAPTER 11 PROBLEMS

Problem 11.1   Explain the concept of an ensemble.

Problem 11.2   Explain qualitatively why the canonical partition function relative to the microcanonical partition function is different for distinguishable and indistinguishable particles as expressed in Equations 11.5 and 11.6.

19.12   CHAPTER 12 PROBLEMS

Problem 12.1   Discuss the importance of the relative fluctuations being proportional to the inverse of the average number of molecules in a specific open volume.

Problem 12.2   Show that $\left\langle (x - \langle x \rangle)^2 \right\rangle = \langle x^2 \rangle - \langle x \rangle^2$.

*Problem 12.3   Derive Equation 12.5.

*Problem 12.4   Derive Equation 12.12.

*Problem 12.5   Show that Equation 12.13 is valid.

19.13   CHAPTER 13 PROBLEMS

Problem 13.1   Consider a 1-L container with 1 $mole$ of neon gas at 25 $°C$. Calculate the collision frequency, the average time between collisions, and the mean free path.

Problem 13.2   Derive Fick's second law from Fick's first law.

Problem 13.3   Compare the diffusion coefficient for spherical particles of radius 1 nm, 10 nm, and 100 nm moving in water at room temperature.

Problem 13.4   Derive Equation 13.15.

Problem 13.5   Consider the first order decay of a molecule by four parallel processes in which the rate constants are $k_4 = 3k_1$; $k_3 = 2k_1$; and $k_2 = k_1$. Calculate the relative yield of each of the four products.

Problem 13.6   Under which circumstances will a pressure perturbation not allow you to measure the kinetics of a reaction?

Problem 13.7   Consider a measuring volume with dimensions of 1 μm × 1 μm × 3 μm. At what concentration will there, on the average, be 9 molecules in that volume? What is the expected average relative fluctuation amplitude for this system?

Problem 13.8   What conditions are required for a measurement technique to be a suitable candidate for fluctuation analysis to be useful in estimating concentrations?

Problem 13.9   Show that Equation 13.36 is correct (compare with Problem 12.2).

Problem 13.10   Derive Equation 13.45.

19.14   CHAPTER 14 PROBLEMS

Problem 14.1   Describe the principle of orthonormality of wave functions and compare this to the orthonormal properties of unit vectors in a Cartesian coordinate system.

Problem 14.2   Define the transition dipole moment. How does this compare to a dipole moment of a molecule?

*Problem 14.3   Prove that the integral of an odd function is zero.

Problem 14.4   What is meant by stimulated emission?

Problem 14.5   Describe the differences between homogeneously and inhomogeneously broadened lines. What is the source of homogeneous line broadening.

19.15   CHAPTER 15 PROBLEMS

Problem 15.1   What is the Frank Condon principle and why is it important in the concept of absorption and emission of electromagnetic radiation?

Problem 15.2   Compare internal conversions to intersystem crossings.

Problem 15.3   Sketch an example of a Jablonski diagram with one singlet and one triplet state.

Problem 15.4   Compare the properties of fluorescence and phosphorescence.

Problem 15.5   Show why the Stern Volmer plot is linear and explain the significance of the intercept and the slope.

Problem 15.6   What are the key differences between the dynamic and static quenching processes?

Problem 15.7   Discuss the possible mechanisms whereby energy can be transferred from one molecule to another.

Problem 15.8   Why is triplet-triplet annihilation critical to photodynamic therapy?

Problem 15.9 What are the three key factors that affect the efficiency of energy transfer by the Förster mechanism?

Problem 15.10 What is the separation of two chromophores whose energy transfer efficiency is fifty percent?

19.16 CHAPTER 16 PROBLEMS

Problem 16.1 What is the key difference between a normal epifluorescence microscopy and a confocal microscope?

Problem 16.2 What is the point spread function for a confocal microscope?

Problem 16.3 Discuss the diffraction limit and the impact on confocal microscopy.

Problem 16.4 Describe two different methods that can be used to circumvent the diffraction limit in super-resolution microscopy.

Problem 16.5 Compare and contrast the decay of the fluorescence correlation function for a system that is diffusing rapidly with that for a system flowing rapidly.

19.17 CHAPTER 17 PROBLEMS

*Problem 17.1 Derive Equation 17.7 from the Maxwell Equations.

*Problem 17.2 Derive Equation 17.11 from the Maxwell Equations.

*Problem 17.3 Derive Equation 17.12 from the Maxwell Equations.

*Problem 17.4 Derive Equation 17.13.

Problem 17.5 Use a spreadsheet or other programming tool to calculate the functions in Equations 17.19 and 17.23 and to plot them so illustrate the subtle differences in the absorption curves.

Problem 17.6 Use Equations 17.16 and 17.25 to show that $I_n = nI_0$.

Problem 17.7   Draw the wave-fronts corresponding to the refraction of a wave that enters a medium of lower refractive index and compare to Figure 17.9.

Problem 17.8   Use the geometry of refracted beams to prove Snell's Law.

Problem 17.9   Describe the difference between s- and p-polarization.

Problem 17.10   What is the Brewster Angle and why it is only relevant to p-polarizations?

Problem 17.11   What is the critical angle and under which circumstances is it important?

Problem 17.12   Describe the appearance of an evanescent wave in the low refractive index medium.

Problem 17.13   Write the expression for the wave equation at the surface of a thin optical fiber.

CHAPTER 20

References

The references are listed by the Chapter in which they first appear and in the order at which they appear.

20.1 CHAPTER 1

Woodrow Wilson International Center for Scholars
(http://www.nanotechproject.org/)
National Nanotechnology Initiative: *Nanotechnology Research Directions for Societal Needs in 2020. Retrospective and Outlook* www.wtec.org/nano2/

20.2 CHAPTER 2

Donald A. McQuarrie and John D. Simon: *Physical Chemistry - a Molecular Approach* University Science Books, Sausalito Ca (1997)

20.3 CHAPTER 4

Herbert L. Strauss *Quantum Mechanics: An Introduction* Prentice Hall Englewood–Cliffs, NJ 1968
http://www.falstad.com/qm1d/
http://www.falstad.com/qm2dbox/
Erwin Kreyszig *Advanced Engineering Mathematics* John Wiley and Sons, Inc New York 1967
Carl. W. David University of Connecticut Chemistry Educational Material
http://digitalcommons.uconn.edu/cgi/viewcontent.cgi?article
=1013&context=chem_educ
http://en.wikipedia.org/wiki/Vibrations_of_a_circular_membrane

20.4   CHAPTER 5

N. S. Bayliss *The Free-electron Approximation for conjugated compounds* Quarterly Reviews, 319-339 (1952)

Prathap Haridoss lectures at IIT Madras available at http://nptel.iitm.ac.in

R.J. Martin-Palma and A. Lakhtakia *Nanotechnolgy: A Crash Course* SPIE Press Bellingham, Washington USA 2010 (SPIE Digital Library eISBN 9780819480767; doi 10.1117/3.853406)

P. Harrison *Quantum Wells, Wires and Dots: Theoretical and Computational Physics of Semiconductor Nanostructures* Wiley 2009 ISBN 9780470770986)

H. Yu, J. Li, R.A. Loomis, L-W. Wang, and W.E. Buhro Nature Materials Vol 2, 517-520 (2003)

G. Ledoux, O. Guillois, D. Porterat, C. Reynaud, F. Huisken, B. Kohn, and V. Paillard Physical Review B Vol 62, 15942-15951 (2000)

V. N. Soleviev, A. Eichhöfer, D. Fenske, and U. Banin J. Am. Chem. Soc. Vol 122, 2673-2674 (2000)

A.C.A. Silva, S.L.V de Deus, M.J.B. Silva, and N.O. Dantas Sensors and Actuators B Vol 191, 108-114 (2014) Y. Suzuki, C.N. Roy, W. Promjunyakul, B. Vasudevanpillai, N. Ohuchi, M. Kanzaki, H. Higuchi, and M. Kaku Molecular and Cellular Biology Vol 33, 3036-3049 (2013)

L.A. Bentolila Y. Ebenstein, and S. Weiss J. Nuclear Medicine Vol 50,493-496 (2009)

V. Biju, S. Mundayoor, R. V. Omkumar, A. Anas, and M. Ishikawa Biotechnology Advances Vol 28, 199-213 (2009)

20.5   CHAPTER 6

P.W. Atkins *Physical Chemistry* 3rd Edition Oxford University Press ISBN 0-7167-1749-2

(or newer editions such as PW Atkins and J. De Paula, Oxford New York, 10th Edition 2014)

Erwin Kreyszig *Advanced Engineering Mathematics* John Wiley and Sons, Inc New York 1967

20.6   CHAPTER 7

H. H. Farrell and C. D. Van Siclen 1441-,J. Vac. Sci. Technol. B Vol 25 1441-1447 (2007)

J. Du, R. Zhao, and Y. Xue J. Chem. Thermodynamics Vol 45, 48-52 (2012)

C. Binns *Introduction to Nanoscience and Nanotechnology* John Wiley and Sons 2010 (ISBN 9780471776475)

N. Lopz, T. V. W. Janssens, B. S. Clausen, Y. Xu, M. Mavrikakis, T. Bligaard, and J. K. Noerskov Journal of Catalysis Vol 223, 232-235 (2004)

20.7 CHAPTER 8

Dr. Rudy Schlaff Tutorials at
http://rsl.eng.usf.edu/Pages/Tutorials/TutorialsWorkFunction.pdf
L Zhou and M. R. Zachariah Chemical Physics Letters Vol 525-526, 77-81
(2012)

20.8 CHAPTER 9

M. Ruth and J.N. Israelachvili Chapter 29 Springer Handbook of Nano-
technology
ISBN 978-3-642-02525-9 (Online)
D. Leckband and J. N. Israelachvili Quart. Rev. Biophys. 34, 105-267 (2001)
Wen Zhang *Nanoparticle Aggregation: Principles and Modeling* Chapter 2 in
Nanomaterial Impact on Cell Biology and Medicine Edited by D.G. Capro
and Y. Chen Springer 2014
N. Li, L. Yu, and J. Zou Journal of Laboratory Automation Vol 19, 82-90
(2014)
H. Ohshima J. Colloid and Interface Science Vol 168, 269-271 (1994)

20.9 CHAPTER 10

Webster's Encyclopedic Dictionary of the English Language (Canadian Edition, 1988)
Weisstein, Eric W. "Continuous Distribution."
From MathWorld–A Wolfram Web Resource.
http://mathworld.wolfram.com/ContinuousDistribution.html
http://startrek.com/lesson2/poisson.aspx

20.10 CHAPTER 13

E. L .Elson and W.W. Webb Ann. Rev. Biophys. Bioeng. Vol 4, 311-334 (1975)

20.11 CHAPTER 16

Y. Liao, Y. Shen, L. Qiao, D. Chen, Y. Cheng, K. Sugioka, and K. Midorikawa
Optics Letters Vol 38, 187-189 (2013))
X. Yu, Q. Bian, Z Chang, P.B. Corkum, and S. Lei Optics Express Vol 21,
24185-24190 (2013)
E. L. Elson Biophysics Journal Vol 101, 2855-2870 (2011)
D. Magde, E. Elson, and W.W. Webb Physical Review Letters Vol 29, 705-708
(1972)

E. L. Elson and D. Magde Biopolymers Vol 13, 1-27 (1974)

D. Magde, E.L. Elson, and W. W. Webb Biopolymers Vol 13, 29-61 (1974)

F. Fujii, M. Horiuchi, M. Ueno, H. Sakat, I. Nagao, M. Tamura, and M. Kinjo Analytical Biochemistry Vol 370, 131-141 (2007)

N.O. Petersen, P.L. Höddelius, P.W. Wiseman, O. Seger, and K.-E. Magnusson Biophysical Journal Vol 65, 1135-1146 (1993)

N. O. Petersen, C. Brown, A. Kaminski, J. Rocheleau, M. Srivastava, and P.W. Wiseman Faraday Discussions Vol 111, 289-305 (1998)

20.12   CHAPTER 17

D. Mittleman Introduction to Waves and Photonics Lectures for ELEC 262 at Rice University http://www-ece.rice.edu/ daniel/262/262home.html

Index